Wright Brothers

An Appraisal and Flying With the Wright Brothers

(The Inspirational Life Story of Wilbur and Orville Wright)

Brian Dawkins

Published By **Bengion Cosalas**

Brian Dawkins

All Rights Reserved

Wright Brothers: An Appraisal and Flying With the Wright Brothers (The Inspirational Life Story of Wilbur and Orville Wright)

ISBN 978-1-77485-683-3

No part of this guidebook shall be reproduced in any form without permission in writing from the publisher except in the case of brief quotations embodied in critical articles or reviews.

Legal & Disclaimer

The information contained in this ebook is not designed to replace or take the place of any form of medicine or professional medical advice. The information in this ebook has been provided for educational & entertainment purposes only.

The information contained in this book has been compiled from sources deemed reliable, and it is accurate to the best of the Author's knowledge; however, the Author cannot guarantee its accuracy and validity and cannot be held liable for any errors or omissions. Changes are periodically made to this book. You must consult your doctor or get professional medical advice before using any of the suggested remedies, techniques, or information in this book.

Upon using the information contained in this book, you agree to hold harmless the Author from and against any damages, costs, and expenses, including any legal fees potentially resulting from the application of any of the information provided by this guide. This disclaimer applies to any damages or injury caused by the use and application, whether directly or indirectly, of any advice or information presented, whether for breach of contract, tort, negligence, personal injury, criminal intent, or under any other cause of action.

You agree to accept all risks of using the information presented inside this book. You need to consult a professional medical practitioner in order to ensure you are both able and healthy enough to participate in this program.

Table Of Contents

Chapter 1: The Early Days: Childhood, Family, And Business _____ 1

Chapter 2: The First Thought Of Flight _____ 43

Chapter 3: A Failure As Well As The Success Story _____ 62

Chapter 4: The Last Years _____ 85

Chapter 5: "A Dream To Fly" _____ 137

Chapter 6: A History Of The First Powered Airplane _____ 158

Chapter 1: The Early Days: Childhood, Family,

And Business

Orville or Wilbur brothers were American brothers, inventors as well as aviation pioneers. Hey are thought to be the creators and designers of the world's first commercially successful airplane, establishing the first powered controlled and sustained higher-than-air human flight. The Wright brothers also designed their flying machine to be the first fixed-wing aircraft that was practical. Wright brothers Wright Brothers were among the very first to develop controls for aircrafts, bringing to flying fixed-wing aircrafts.

Another remarkable invention by them was the three-axis control , which gives the pilot the ability to effectively steer the aircraft and maintain equilibrium that is crucial. This technique became standard for fixed-wing aircrafts of all types.

From the beginning that was they Wright brothers' primary goal was to devise a robust method for pilot control which was the primary factor in solving "the flying issue". This method was a totally innovative one and markedly different from other pioneers in the era, who focused on the development of powerful engines.

Through the use of an unassuming wind tunnel built at home The Wright brothers gathered more accurate information than any before. The data they collected allowed them to create and build propellers and wings which are more efficient than the ones created before.

The very the first U.S. patent was not the invention of the flying machine, it was more of an creation of a system of aerodynamic control, which was used to manipulate the the surfaces of a flying machine.

A significant influence on Wright brothers was their working years in their shop using motors, printing presses bicycles, and various other equipment. The experience was significant influence on their professional capabilities. The tasks they did on bicycles, helped shape their conviction that a car that was unstable like a flying vehicle could be managed and balanced through training.

In the years 1900, they were engaged in conducting lengthy glider tests, which also helped them develop their skills as pilots. One of their bicycle shop's employees Charlie Taylor, a bicycle shop employee Charlie Taylor, became an integral part of the team later and helped build their first airplane engine.

It is known that the Wright brothers' reputation as creators of the plane has been contested by different parties. There are a myriad of debates over the various different claims made by early aviation pioneers. The historian Edward Roach argues that they were highly self-taught engineers however, they were poor businessmen.

It was believed that the Wright family was among the oldest families in Ohio; Wilbur and Orville's grandfather was the one who helped settle Miami County. Due to their father's work it was not uncommon for them to travel a lot throughout their childhood. However, eventually, they settled to take a rest at Dayton, Ohio.

Milton Wright and Susan Catherine Koerner Wright were the parents of Wilbur and Orville. They got married in 1859. The Wright family is a huge one. Wilbur as well as Orville had two brothers and a sister. Reuchlin Wright. Lorin Wright, and Katharine Wright.

Parents

Milton Wright and Susan Catherine Koerner Wright were married in 1859. Milton Wright was born in 1828 on the farm of his father, Rush County, IN. He was farmer until when he joined in the Church of the United Brethren in 1847. In 1856, he was ordained as a minister in the church. When he was enrolled at Hartsville College in 1853, He was introduced to Susan Koerner. They were married in 1859, and were blessed with seven kids between the years 1861 between 1861 and 1874. In his lifetime, Milton served as a circuit-riding minister for the United Brethren Church, an editor of the Religious Telescope, a professor of theology, and a elected Bishop of his church. His work in the United Brethren Church made him to move around with his entire family frequently.

The Wrights resided throughout Dayton, OH, Cedar Rapids, IA, and throughout Indiana. His job required travel up to 8000 miles in one year. While traveling, Milton was in close contact with his family and wife and wrote hundreds of messages to them. Milton was known as an kind and caring father. Milton and Susan supported their children's curiosity, and had two large libraries in their home for them to access. There were occasions that Milton permitted his children, even his daughter, to take days away from school to pursue their own pursuits. It was in 1889 that Milton Wright broke with the liberal direction within the United Brethren Church. In 1889, he founded his own sect of conservatives, called the Church of the United Brethren, Old Constitution. Milton was laid off from all of his positions when he was 77 years old. He died in 1917.

Susan Wright was born in 1831 within Loudoun County near Hillsboro, VA. Her father, John Gottlieb Koerner was a skilled wagon maker. He emigrated from Germany to America from the small village in Forthen, Germany in 1818. He was a car manufacturer in Baltimore for a time before he was married Catherine Fry in 1820. After 1832, when Susan was born the Koerners relocated to a farm with a size of 170-acres Union County, Indiana south of the town of Liberty. Susan despite being female, was able to operate a variety of equipment used on the farm with a great deal of skill. Her family members recalled her as an expert in mechanical work and extremely hand-skillful. Susan was admitted to the U.S. Brethren religion in 1845 at the age of fourteen.

When Susan was 22 when she was 22, she enrolled at Hartville College in Indiana. There she excelled in sciences and literature. Milton Susan and Milton Susan got married on 18th May,1859. Susan gave him seven kids between the years 1861 and 1874 however, two of them passed away shortly after the birth.

The year 1883 was period that the family lived close to Richmond, Indiana, Susan began to exhibit symptoms from "the consumption" tuberculosis. Susan was totally disabled at the time of her death in 1889.

Siblings

Wilbur Orville and Wilbur Orville were the sixth and third born of seven children.

Reuchlin Wright was born 1861. He was the eldest child of the Wright family. He attended Western College near Cedar Rapids, IA in 1879, and later taught for a short time in elementary school. He then spent a period studying at Hartville College, IN with his brother Lorin then was employed as a clerk at the woodyard located situated in Dayton, OH. He married Lulu Billheimer. He was the son of United Brethren missionaries. They had an infant daughter in 1886 called Catherine Louise, a year after their wedding.

He was born in 1901 and relocated to a farm close to Tonganoxie, Kansas, where he raised cattle and corn. The first child he had died shortly after however Reuch and Lulu had three additional children, including Helen Margaret, Herbert, and Bertha Ellwyn. Reuch died in 1920. Many historians believe Reuch as the "black sheep" of the Wright family. It was believed that he was at war along with the father Milton. There was likely to be tension between Milton and Reuch. Even though Reuch was located far away from Dayton, Milton visited him whenever was possible. Reuch also went to Dayton and his correspondence with his family members was warm and friendly. Reuch was frequently very negative of himself.

Lorin Wright was born 1862. He spent time in the Kansas frontier and later went to Hartville College, for a year in 1882. He was employed as a bookkeeper at an establishment selling carpets located in Dayton, OH and courted Ivonette Stokes. Lorin and Ivonette his wife were married in 1892. They had four children: Milton, Ivonette, Leontine and Horace.

The year 1900 was when Lorin was able to help Katharine the younger sibling of his, run his sister's role in the Wright Cycle company while their brothers were working in Kitty Hawk, NC. He also began his own "street sprinkler" business to earn an extra income.

He went to visit Wilbur as well as Orville on the way to Kitty Hawk in 1902, photographing their glide experiments at Kitty Hawk. Lorin was always friendly and helpful to his siblings and especially Wilbur and Orville who needed an ample space to conduct propeller tests. Lorin offered them his barn for carriages. Lorin played a major role and was a key factor in the success of his brothers. The year 1911 was the time that Lorin as well as his son Horace went along with Orville To Kitty Hawk with a new glider. Lorin was able to assist his brother in helping him achieve the world's first record for speed.

Following the time that Orville transferred The Wright Company, Lorin bought an interest in Miami Wood Specialties, a company that made toys, including those called Flips and Flops that Orville had created. Lorin died in 1939.

Wilbur Wright was born on 1867. He was a great athletic and academic performer. He earned graduation from high school during his time at Richmond High School located in Richmond, IN, but He never sought an official certificate. The primary reason could be due to the fact that his family relocated from Dayton, OH, just prior to his graduation. He graduated in 1885 and enrolled in some college preparatory classes within Central

High School in Dayton, OH, He was extremely ambitious and wanted to go to Yale university, but he never was able to attend the college. He stayed home, and cared for his mother's illness until her death tragically occurred in 1889.

Later his younger brother Orville joined him in the business of newspapers as the editor of The West Side News and later, The Evening Item. Following the collapse of the newspaper businesses, Wilbur became a partner with Orville in a printing firm as well as a repair shop for bicycles as well as a company that made bicycles.

The year 1896 was a significant one for everyone because that was the year Wilbur and Orville began to become fascinated by aviation. They began their first aeronautical experiment using kits in 1899. They constructed a number of gliders until 1902 and then developed an aerodynamic control system to aircrafts, while also learning to fly. They added engines to their plane in 1903 before completing the first sustained, controlled powered flight on the 17th of December of that year. They continued to improve their equipment as well as their ideas.

The year 1908 was the first time they offered airplanes in 1908 to US Army and to a French syndicate, and then demonstrated the aircraft to the public. He passed away in Dayton on the 30th of May 1912, exactly 13 years after the date he started the first of his formal flight research.

The twins Otis as well as Ida Wright were born in 1870, but unfortunately passed away when they were infants.

Orville Wright was born on 1871. Orville was a great pupil in his elementary school days, but when he grew older and pursued different interests his academic performance suffered. He was a fan of making woodcuts and printing them. In 1889, after his mother passed away and he was unable to go back to the senior year of high school. Instead, he decided to print the newspaper he had created, The West Side News and later in The Evening Item; however the newspaper failed in the same business later. In 1896, the Wrights brothers began to become fascinated by aviation. They conducted the first aeronautical experiment using kites in 1899 and they built a succession of gliders from 1902 onwards They designed an aerodynamic control method for aircraft while learning to fly.

The Wright brothers formed their Wright Company to manufacture airplanes in 1909 With Wilbur as president. Wilbur passed away from typhoid 1912, following which Orville reluctantly assumed the presidency. He sold the business in 1915 and resigned to pursue his own passions. Following that, he served as a board member for life on the National Advisory Council on Aeronautics (NACA) which was later changed to NASA The National Air and Space Administration. Orville passed away in 1948.

Katharine Wright was born in 1874. Her birthday was shared along with Orville Wright. She took over her mother's position as the head of the Wright family, after turning 15 years old, and continued to be the lover at Wright's Wright house until the year 1926. She was the sole Wright child who graduated from college. She earned her degree in teaching in 1898 from Oberlin College in 1898. She was a professor of literature of classical origins and literature at Steele High School in Dayton, OH. When Orville was seriously injured in an accident on the air in 1908 she cared for his recovery, and then traveled to France together with him to join their brother Wilbur and flew together with Wilbur to France to the very first time.

After Orville Wright sold his Wright Company in 1915, she stayed with Orville until 1926. At that time, she was introduced to her college friend Henry Haskell and a sudden romance blossomed between them. They had plans to marry however, this thought enraged Orville because he couldn't imagine living without Katharine and was reluctant to attend at the reception. Henry and Katharine resided together in Kansas City, where he was editor for The Star newspaper. After their wedding, two years later, Katharine contracted pneumonia. In her final days, Orville travelled to Kansas City and was there with Katharine when she passed away in 1929.

Humans dream of flying

The first humans observed birds flying, to now, every one of us would like to fly in the same way as these amazing creatures can. Since the beginning of our human history there were many who tried to fly. They tried to learn the mechanisms that allowed them to do this. They observed the bird and wanted to replicate their skills. It's always been that was impossible, inaccessible, quite distinct and far away.

Before the Wright brothers, there were many remarkable people who had tried to fly, contracted engines and other machines to make their dream to fly come true.

"The dream of flying is a concept handed from our forefathers, who during their long, arduous journeys across untracked lands in prehistoric times observed with envy the birds that flew freely through space in full speed, above all obstacles in the endless highway of the air." Wilbur Wright. Wilbur Wright

Let's examine the human story and pay tribute to the many outstanding individuals who longed to fly.

Daedalus and Icarus

One of the oldest tales of flying originates from the Greek mythology that tells the story of Daedalus as well as his brother Icarus. They longed to escape the wrath of in the hands of the king Minos who was on Crete and flew using wings made from feathers that were held by wax. Daedalus advised his son to not fly close to the sun. However, he did not listen to his advice. Thus, the wax on his wings began to melt and he sank into the ocean and drowned.

Leonardo da Vinci

This extremely talented Italian was really keen to fly. Initial attempts to fly were made by attempting to imitate the flight of birds' flapping wings however human hands are very inflexible to flapping wings for long periods of time, so machines were developed to help legs or arms to do the flapping. These machines are known in the field of "ornithopters."

Leonardo da Vinci was trying to develop and build a machine during the early 1480s up to the time of his death in 1519. But he did not build one of his machines so far as we know to us. He wasn't able to fly as his designs were not aerodynamic characteristics.

Sir George Cayley

Cayley was an engineer and baronet who was a resident of the estate of Yorkshire, England, was the first person to develop the idea of the modern aircraft. He wrote "The entire problem can be solved within these limitations - to construct a surface to be able to support a weight given by the applying force against the resistance of air."

He wrote three articles in 1809-10 about his aeronautical research called "Aerial navigation." He created an early model glider with an upward-sloping front fixed wing, and an elongating tail.

In 1853, when he was at 80 years old, the man designed an all-size glider which carried his hesitant coachman on the air across a small valley.

William Henson

Henson tried to apply Cayley's ideas in the design of an airplane that could be powered by steam engines. The name was "Aerial Steam Carriage". He received his patent for the design in 1842. His invention was the first to allow airscrews for powering the fixed-wing monoplane. The bracing and structural design prefigured modern designs.

John Stringfellow

Then in 1848 another dreamer named Stringfellow created a more refined version. He began by dragging it down an inclined wire of 33 feet. He then let it go while the engine was running.

In general, the model demonstrated real powered flight by climbing slightly before it hit an obstruction.

Steam engines were too heavy to support the power it generated.

Alphonse Penaud

Penaud is a Frenchman and was among the very first person to make use of twisted rubber bands

motor power in a model helicopter. The helicopter could rise up to the ceiling when operated.

The year was 1876 and he was able to patent an airplane design exactly like modern aircraft. The plan included an "joy-stick" that was designed at controlling vertical and horizontal Rudders. It was a design feature which was in the process of predicting controls. When he faced financial challenges to building his plane He was depressed and took his own life at the age of thirty.

Francis Wenham

An Englishman Wenham created, constructed and utilized the very first wind tunnel in 1871. His tunnel was a large wooden box, with steam-driven fans at the end.

He advocated using multiple wings over each other. He was also the first to obtain the first patent for the first flying machine made of superposed planes.

Thus, he was known for being"the "father of biplanes" an idea later adopted in the Wrights.

Otto Lilienthal

Lilienthal invented and flew the first gliders that were successful in the history of gliders. They were similar to hang gliders of today. The pilot began his flying journey in 1867, when he was a child in Germany. With the assistance of his brother Gustav He built a number of small gliders and flew controlled flights using these gliders.

Lilienthal was successful in the development of his innovations, and during the time period between 1891 and 1996 He was able to make more than 2500 successful glider flights.

As he glided on August 9 in 1896, he was struck by a powerful breeze that caused his glider to be tossed upwards to 50 feet, at the most acute angles. Lilienthal passed away the following day from a fractured back at age 48.

Octave Chanute

Chanute was an entrepreneur who was also a civil engineer as well as railroad bridge construction expert. He gained notoriety after his middle age when he began to take an interest in aviation. He flew with multi-wing gliders near the banks of Lake Michigan in 1896, seeking a design that could provide stability on the fly.

In 1894, he released "Progress on Flight Machinery." The book was regarded as to be the primary reference for anyone who was interested in flying.

Wilbur wrote in Chanute on the 13th of May 1900 "For certain years I was plagued by the idea that flying is possible for humans." This was the start of a 10-year-long relationship among Chanute as well as Wilbur.

Childhood

Orville and Wilbur Wright were born 4 years apart in different cities. Both were curious and shared their curiosity to the world at large. Wilbur had been born around 1867 on an unassuming farm in Millville, Indiana. Orville came into the world in 1871, in a house situated in Dayton, Ohio.

The life in the Wright family was strict but also loving. The parents advised their children to do what they liked and whatever they wanted to. They had a huge library of books covering various subjects. Orville and Wilbur were both avid learners since the very beginning of their in their childhood.

Orville and Wilbur's got attracted to flying after they received a present from their father. It was a toy that flew. It had a body made of paper as well as other components comprised of bamboo and cork. The boys loved the present so much that they were enthralled by it. They were aged 7 and 11 when they received the present and decided to make it themselves. They recalled how it was and promised that they would one day fly through the air just as the toy.

They were fascinated by flying and mechanical objects. Orville sold kites to school to earn money, and Wilbur was reading as much as the information he could on how birds fly and how machines functioned. While they were great students, neither completed high school.

Wilbur was struck on the face by an baseball bat as a teen. He was afflicted with irregular heartbeats for throughout his life. He took care of his mother who had tuberculosis. This disease at the time was a total disaster for many, as it was not curable.

Orville quit high school. He was planning to establish an enterprise of printing. He along with Wilbur created a printing press that was very successful. Later, they sold their printing businesses. They were excellent mechanics and they could fix anything that anyone handed them. They learned this from their mother who was their family mechanic.

In the bicycle shop, the concept of an airplane was birthed. The Wright brothers built kits, particularly massive ones. They were referred to as gliders. Orville and Wilbur constructed them by their own. They built the gliders big enough to let a person ride on. They flew using only air flow. The passenger could take around 10 seconds before the glider slid to a stop.

Printing

The Wright Brothers had two earlier enterprises prior to the launch of their aircraft business. The first being editors, writers publishers, printers and editors and the second one was bicycle makers and dealers of bicycles.

In their early days of printing, Wilbur and Orville wrote edited, published and edited fifty issues in a week-long paper, "The West Side News" and 78 issues of the daily newspapercalled "The Evening Item." Additionally, the two printed hundreds jobs.

Orville began his printing career at 15 in 1886. He along with his pal Ed Sines; he owned a small printing business and produced their first newspaper "The Midget" for their classmates at school. The newspaper was supposed to be an annual publication, but it ended up being not a success. It was only for one issue because the father of Orville Milton was dissatisfied with the efforts.

The issue in the document was printer brothers left the 3rd page blank , aside from their names, Sines and Wright. They were exhausted, since each page was printed in a separate manner and the type was made by hand. The Mr. Wright Milton once told themthat "They did not do themselves justice by slapping the 3rd page." People reading the text could "get the impression that they were inefficient and unresponsive."

Ed as well as Orville continued to print in the business. They started their printing at Ed's but the business was booming enough to allow them to purchase an additional press and moved to a different location, which was behind the Wright home located on Hawthorne Street. On cold winter days, they performed their typesetting inside their home. They even enlisted a neighbor boy to assist them at 15 cents per week.

But they were unable to remain in place for very long with no disagreement. Thus it was that it was decided that the "Sines and Wright" business arrangement changed following an argument regarding what to do about the pop corn they were paid for. Orville wanted to purchase a larger varieties, whereas Sines wanted to take the popcorn. The disagreement was settled by Orville buying the Sines shares and Sines accepted to remain for Orville. The arrangement was in place throughout the Wrights printing business, which was eventually sold in 1899.

The two Orville as well as Wilbur were introduced to the printing industry from a young age. They were able to inherit it from their father, a bishop of the United Brethren in Christ church as well as a writer of religious inclination editor, publisher and writer.

In 1869, he was chosen to be the editor of the church's magazine, "The Religious Telescope." Since the job required for it, the entire family relocated from Dayton, Ohio to Dayton, Ohio where the church owned a massive printing facility located in the center of the city center Dayton. Milton was frequently seen by his children at the printing facility. In particular, Orville especially was thrilled by the steam-powered printing presses.

Orville was the very first kid to become obsessed with an even bigger press that he could use. The spring of 1888 at the age of 16 and with the help of his older brother Wilbur and his brother Wilbur, he constructed an industrial printing press using the fold-up top of Katharine's baby buggy the remains of a tombstone that he used as an iron bed along with firewood, and other scrap pieces from an abandoned junkyard. After a couple of weeks the brothers' machine was producing 1,000 sheets per hour. The brothers were proficient in the field of mechanical work.

Orville developed his understanding and abilities in the field of printing while working two summers at an local printing company, when he was only 15and 16 years old age. He quit high school before his senior year so that he could devote all of his energy and time to the job that he enjoyed.

In the month of March, 1889, Orville started printing and publishing a weekly newspapercalled The "West Side News." It featured three columns with four separate pages. The price for subscription was 40 cents for the year or 10 cents per 10 weeks. The newspaper did quite well and Orville moved to a smaller office at the 1210 West Third Street. The paper grew from three columns to four columns. In the following years, Wilbur was also joined by his brother. Wilbur was editor, and Orville the editor. Orville was the.

Wilbur was a student at Yale and a career in teaching as a goal after high school, but an accident that was unfortunate threw off all of his plans. While playing ice hockey in the cold winter months of 1885 Wilbur was struck in the mouth by a stick from a hockey. The impact smashed off several of his teeth. The psychological and physical impact on Wilbur was overwhelming. Wilbur also suffered from stomach problems and heart palpitations. There was a fear that permanent injury could result. The treatment was a prolonged time of relaxation.

At the end of 1886, his physical ailments appeared to be gone However, he suffered from persistent depression. Wilbur was devoted to caring for his mother when she was ill.

The first documents that was published under the name of Wilbur, was a brief church tract known as "Scenes in the Church Commission on the Last Day of its Session." It was published in 1888.

In April 1890 In 1890, the Wrights created a brand new newspaper that was called "The Evening News." The paper featured five columns with over half of them containing international and national news. Also, there were baseball scores that were from both The American Association and the National League.

The July 17th and June 26th issues of "Item" featured articles about the activities of the renowned German glider researcher Otto Lilienthal. His work and research had a an impact on the Wright brothers in their future aviation business. "Item" was published in 1890 "Item" ended at the end of July, 1890, after just 4 months since its publication. The newspaper moved in late 1890 to the newly constructed Hoover Block at the corner of West Third and Williams St.

In this particular location, they produced a newspaper oriented towards black, "Dayton Tattler," as well as additional printing projects in the name of Paul Laurence Dunbar, well-known black poet.

Then, in 1892 , the brothers began to become interested in bicycles and set up their first bicycle store at 1005 West Third St. The shop was a place for services and sales.

They continued to print. They printed a magazine in 1894 titled "Snap-Shots of Current Events." The document included a number of articles on bicycles as well as essays, and jokes.

In February 1896 they reduced their publication to "Snap-Shots" and moved it to the 22nd floor of South Williams St. They identified Wright Cycle Co., as the publisher of their magazine. The printing department was located in the upper floor while the bicycle company was located at the top of the building. It was the very first occasion two businesses shared a space.

In 1897, the brothers moved their printing and bicycle businesses in 1127 West Third St. It was this building which gliders and Wright Flyer were conceived and constructed. The printing facility was located situated on the 2nd floor.

Orville has never stopped his love for printing. In 1930 he designed and constructed an printing press for Miami Wood Specialty Co.

Bicycles

Before they made aircrafts The Wright brothers made bicycles. This was their second venture following printing. Similar to many Americans in the 1890s, they decided to run a bicycle-based business. This type of bicycle was much simpler to ride and mount as compared to the "ordinary bike."

First bicycles

Wilbur Wright purchased a second-hand ordinary high-wheel bicycle at a price of just $3 during the time when the Wright family resided in Richmond between 1881 to 1884. The year 1892 was when Orville purchased a brand newly-built Columbia security bicycle that cost $160. The same year, Wilbur bought a secondhand Eagle security bicycle at $80. The two of them were really into cycling and Orville was racing at times. He was once awarded the rocking chair at the race during the Montgomery County Fair.

First bike shops

The Wrights set up an repair and sales shop, dubbed The Wright Cycle Exchange located at 1005 West Third Street in Dayton, OH in 1892. There were a wide variety of bicycles, such as Fleetwing, Reading, Coventry Cross, Envoy, Smalley, Warwick, Duchess, and Halladay-Temple. Prices varied between $40 and up to $100. The

Wrights also provided bicycles for rent as well as sold parts and accessories.

It was not by chance that they chose to open in 1892, or that they selected a location at West Third Street. They were the League of American Wheelmen held their annual meeting for the 12th time at Dayton between July 4 and 5, 1892. It was a major celebration. There were thousands of cyclists who came to Dayton to take part in 13 different races to win prizes upwards of $500. The most significant thing part was the fact that riders were allowed to explore the city. By far the most well-known location to visit was The Central National Soldiers Home located located west of Dayton along Dayton-Eaton Pike. It was also known by the name of West Third Street. Thus, they would travel by the site and visit and use Wright Brother's store.

Locations of bike shops and their names

Since their business was flourishing, Wright brothers moved their bicycle shop six times and changed their name a few times.

1892 1892 Wright Cycle Exchange at 1005 West Third Street, Dayton, OH.

1893 1893 Wright Cycle Exchange at 1015 West Third Street, Dayton, OH.

1893 - 1894 Wright Cycle Exchange at 1034 West Third Street. It was changed later into Wright Cycle Co.

Between 1895 and 1897, Wright Cycle Co. at two locations: the main store located at 22 South Williams Street, Dayton, OH and a branch located in downtown Dayton located at 20 West Second Street. The branch store was shut down in 1896.

From 1897 until 1908 from 1897 to 1908 - 1897 to 1908 - Wright Cycle Co. at 1127 West Third Street, Dayton, OH.

Bicycles manufactured by the manufacturer

In the latter half of 1895 The Wrights decided to make bicycles for themselves. They launched"the "Van Cleve" on April 24th in 1896. In the following year the Wrights presented a different cheaper model, which was dubbed the "St. Clair." The name was derived from local historical events; Arthur St. Clair served as the president in the beginning of the Northwest Territory, which later changed into Ohio, Indiana, and Illinois.

It was the Van Cleve was mostly hand constructed using a variety of handles, wood or metal rims and double or single tube pneumatic tires. The Van Cleve was sold for $65 at the time it was it was first introduced.

The St. Clair was built with high-quality parts that were readily available from a variety of sources like those of the Davis Sewing Machine Company of Dayton along with Pope Manufacturing of Boston, MA. It was offered for sale at $42.50. Between 1896 and 1900, the Wrights produced around 300 bicycles.

There's a bit of confusion over whether Wright's made an automobile called"the "Wright Special." The sole reference to this bicycle was an announcement published on 17 April in 1896 "For several months and years, Wright Cycle Co. Wright Cycle Co. have begun to plan to make bicycles...we will be able to send several prototypes available in the next 10 days or a week, and we'll be ready to take orders by the middle of the month. The WRIGHT Special will be nothing but top quality materials. "

It can be seen from two distinct angles: either the Wrights were ready to unveil their own bicycle or were planning to launch a new bicycle known as"the" Wright Special.

Innovative cycling technologies

The Wright brothers invented two new inventions using their bikes. The Van Cleve had a unique "self-oiling hub." The Wright brothers sprayed the bearings using felt washers and built an oil reservoir in the hub, which cut down on maintenance. The hub was also equipped with spare parts of its own that included two additional race bearings or "cones" inside where the bearings rode. They were among the preferred components to be found on early bicycles.

In 1900 the Wrights displayed the Wrights a "bicycle pedal that could not be undone." The pedals were attached to the crank via threaded posts. Early bicycles had both posts were equipped with normal right-hand threads. As the cyclist pedaled the pedals, it was likely to make one pedal tighter, and loosen the other, and the result was that one pedal would keep falling onto the road. Wilbur and Orville utilized right-hand threads on one post, and left-hand threads on the other, so that the pedaling action would tend to increase the tension on both pedals.

Profits from business

In the beginning, the bicycle industry was a success to initially for the Wright brothers. In their highest year of 1897 they made around $3000, or $1500 at a time that most American worker was able to earn $500 annually. They also saved $5000 which allowed to fund their flight experiments. However, this opportunity was not to last for long. From 1898 onwards it was a time of serious "shakeout" between small bicycle manufacturers , as they shut down or were sold to larger companies. Massive manufacturing companies appeared to produce bicycles, selling them at as low as $10 per piece around the turn in the 19th century. In the years following, Wrights found themselves having to cut their prices repeatedly to keep their prices competitive.

It is vital to point out that during the same year in which the Wright bicycle business was beginning to fall apart (1898) The the Assistant Secretary of The Secretary of the Navy Theodore Roosevelt convinced the War Department to pay Secretary of the Smithsonian Institution Samuel Langley $50,000 to help develop his 1896 Aerodrome into a man-carrying machine. Although this was intended to be kept secret, it was the biggest amount ever given to the War Department to develop a weapon. The news was then became public.

Wilbur And Orville were already looking into the "flight issue" and were aware that all aircrafts at the time were not properly controlled. In the case of Langley Aerodrome had none at any point. In the following year, the brothers began their research on their own with the specific intention of developing an aircraft control system.

Although, they experienced a number of failed attempts, they didn't quit. In the mid-1902 period, following tests of two gliders that failed in the years 1900 and 1901, Wilbur confided to his friend Octave Chanute that the business of bicycles was suffering and he was seeking an alternative manufacturing line. Chanute recommended small-sized heaters to the carriages or electric refrigerators. A few months later, Wright brothers flew their 1902 glider and achieved the aeronautical breakthrough, while keeping their focus on planes.

Selling the business

After 1902 Wrights made a small number of bicycles. There was no bicycles until 1904. They were very involved in their aircrafts in search of an market for their products. When they started selling planes, in 1909, the bike shop located at 1127 West Third Street was converted into a machine store, in which parts for aircraft engines and drive train were offered for sale.

In 1909 or 1910 in 1909 or 1910, the Wright brothers sold all of their remaining bicycle parts as well as the right to use the Van Cleve brand to W.F. Meyers who was an expert in bicycle sales as well as a mechanist and repairman.

Things you don't have any idea concerning the Wright brothers

The two brothers were fascinating and curious. They were excellent editors, mechanics and businessmen. Here are some fascinating details about the Wrights that you may not know about.

1. Coin toss during the first flight

The brothers toss a coin to determine who would be the first to try the Wright Flyer on the sands of Kill Devil Hills, North Carolina. The older brother Wilbur was the winner, however, his first test on the 14th of December 1903 failed and damaged the plane. Three days later Orville lay on his stomach, on the lower wing of the plane and took control of the plane. When he arrived at the airport, around 1035 a.m. The Wright Flyer moved down the guide rail, with Wilbur working alongside to support the delicate aircraft. In just 12 seconds the plane was off the ground, before coming at 120 feet in soft sands. Brothers exchanged their turns in the cockpit for three more times throughout the day.

2. A toy for those who are obsessed with flying

When the brothers were young In the year 1878 their dad came home in the evening with a present dropped in the air. In 1908, they told, "Instead of falling to the floor, as we had expected it flew over the room until it hit the ceiling, and then floated for a time before finally was lowered to the ground." The helicopter was the reason the brothers were so interested in flying.

3. No of them had a diploma from a high school

Wilbur completed four years of high school. However, the family relocated out of Richmond, Indiana, to Dayton, Ohio, before the time he was able to receive his diploma. Orville was a curious kid however, he was kicked out of high school prior to his senior year to begin an printing company.

4. The brothers didn't get married.

The close-knit brothers, who were born four years apart have dedicated their lives to their work. Wilbur said to journalists that he didn't have time to have his wife and an airplane.

5. The Wright brothers flew just once

Orville and Wilbur told that their dad, who was worried about the loss of both boys in an accident involving an airplane that they would not fly in tandem. The father made an exception on May 25 in 1910. He permitted the brothers to take the flight for six minutes near Dayton and Dayton, with Orville piloting the plane and Wilbur as the passenger.

6. The first time it flew in 1903, the Wright Flyer never flew again

The brothers flew four times on the Wright Flyer on the 17th of December 1903 as well as during their flight, a abrupt gust of wind gripped the plane and then flipped it numerous times. The plane suffered severe damages to its ribs, chain guides and motor. It was then beyond repair. This Wright Flyer was returned to Dayton and then never flew again.

7. Orville had been involved in fatal first aviation crash.

After their 1903 success The Wright brothers continued to work on their aircraft development. They introduced their two-person Wright Military

Flyer to the U.S. Army, which needed the demonstration. On the 17th of September 1908 Orville flew into the sky to fly a demonstration at Fort Myer, Virginia, with Army Signal Corps Lieutenant Thomas Selfridge as a participant. Within a couple of minutes during this flight, the propeller abruptly broke up, the plane spiralled out of control , and then it crashed into the ground with a roaring speed. The lieutenant was killed within hours. Orville was in hospital for six weeks following the occurrence of an injury to his leg that was broken with four broken ribs as well as an injury to the back that hampered his mobility throughout his life.

8. Orville was unable to give to the Wright Flyer to the Smithsonian Institution for a long time

In 1903, the Wright Flyer is one of the most sought-after exhibits in the Smithsonian's National Air and Space Museum However, for decades Orville did not want to make a donation to the museum. In 1914 the Smithsonian tried to restore its image of former secretary Samuel Langley, whose work in his Langley Aerodrome failed within nine days of when the Wright brothers took off--by changing the design of the aircraft, and later concluding that it was the one of the first machines "capable" of flying manned. An angry Orville offered to loan his Wright Flyer overseas to the London Science Museum in 1925 in the belief that it was "the only method to correct the past of this flying device that through false and deceitful claims has been misrepresented in the hands of The Smithsonian Institution."

Following the Smithsonian acknowledged it was a misrepresentation in 1940. they was not as it claimed, Orville agreed to donate the plane to the museum. It was eventually donated in to the Smithsonian in 1948, one year following the passing of Orville.

9. Neil Armstrong carried a piece of the Wright Flyer with him to the moon

A second aviation pioneer out of Ohio, Neil Armstrong, was the first person to walk into space in the year 1969. in his spacesuit's pocket was the muslin material that was left from the 1903 Wright Flyer as well as the wood piece from the left propeller of the plane.

Chapter 2: The First Thought Of Flight

From the moment Orville and Wilbur were introduced to the Penaud model helicopter at Cedar Rapids, Iowa, their curiosity about flying and being able build a machine that could fly ignited.

In the early 1890's Both Wilbur and Orville read every article they came across on a science-related topic and discussed the subject. At times, an article in an article that made it directly to the Wright home included a discussion of the attempts to fly by humans. With time the kind of stories attracted the Wright brothers increasing. The year 1895 was the time that two brothers were fascinated by a piece they found during glider research in Germany conducted through Otto Lilienthal. He was gliding in the air, along the slope of a hill on a machine that constructed by him. The two brothers were amazed by Otto and would like to know much more information on Lilienthal as well as his works. They found only a few details about Otto or his work, however, what they did discover sparked a lot of fascination for the two brothers.

In 1896, the summer's enthusiasm for Otto's work not gone. One day, when Orville was in a state of delirium from the illness, Wilbur read that Lilienthal was killed in the crash of his own glider. When Orville was well enough to be able to read the tragic news about Lilienthal's death and the brothers were both in more eagerness than ever before to find out more about Lilienthal did and the accomplishments he had. Brothers were trying to collect all the information they could. They were using everything from the Dayton library.

Knowing that the Smithsonian Institution, at Washington was fascinated by the topic of human flight and flight, they decided to write an inquiry to the Smithsonian soliciting suggestions for reading materials. They got a reply in 1899. These suggestions are made in the book "Octave Chanute's Progress in Flying Machines; Professor Langley's Experiments in Aerodynamics; and the Aeronautical Annuals of 1895, 1896, and 1897 published by James Means, which contained reproductions of the accounts of various experiments, which go from the time of Leonardo da Vinci. Alongside this list of recommended readings as well, the Smithsonian provided a number of pamphlets, which were reprints of materials that they had taken from their annual reports. Included were the works of Mouillard's Empire of the Air, Langley's Story of Experiments in Mechanical Flight as well as an article by Lilienthal on the Problem of Flying and Practical Experiments in the Sky.

The Wright brothers were particularly fascinated by what Otto had accomplished. With hundreds of small flights, he was able to have the most flying experience than any other pilot even though Otto was flying for a total of just five hours over five years. Lilienthal was considered to be the Wright's favorite. They believed they were the only one who did the most in the field of flying. Their research provided a wealth of interesting thoughts, for instance, Wrights started to think about possible solutions to equilibrium problem.

The readings, while expanding their library of information, provided the Wrights plenty of misinformation. One incorrect idea they gathered was that the men had already developed propellers and wings with sufficient efficiency that the motors were able to maintain the machine in the air. Another incorrect idea they had was that the most difficult thing was to keep the equilibrium. They were also deceived into thinking that fore - and the control of an airborne machine was far more difficult than the control of lateral movements.

Orville discovered that nobody, not or even Otto Lilienthal had ever tried the most effective method of ensuring the balance of the lateral. Orville however, he asked; "Why? Isn't it possible to allow the operator to alter the angle of sections of wings towards the tips, and thus gain the force needed for wing tips that are opposite?" Orville had hit on the fundamental premise.

Orville did a sketch of a wing with a stationary section in the center, comprising about one-third of the wings in length from tip to tip with two sections adjustable on each side. These sections were hung via shafts connected by cogs that were positioned on the central section, and extended towards the tips of the wings. The motion of a lever connected on one shaft could result in one wing section to turn in one direction, and the other wing rotated around the reverse. As a result, more lift can be achieved on the side was required.

But, after a few years, they realized the primary two reasons this particular design could not offer a suitable design for a gliding machine. The first is that with two-thirds of the machine's weight and the operator being carried by two shafts structure could be weak. Second, with the wings' ends being able to move freely around to rotate around the shaft, it wouldn't have enough rigidity to support the machine to be carried around.

In the summer of in 1899, Wilbur and his brother made an biplane kite and Wilbur along with a small group of youngsters as observers, flew it over a common near the edges of town. The kite's wing surface was 5 feet wide from top to tip and thirteen inches in width. The surface warping was possible through using four cords that extended across the kite and to the ground. Two cords were tied to the front corners of the tips of the right wing one towards the higher and the other to the lower. The other cords, on their ground level, were tacked on opposite sides of a small stick that was placed in the hands of the operator. The cords attached on the left side of the wing placed in the same manner. With the use of a stick with each of his hands, an user could maneuver the wings however they preferred.

A balance between the front and back was maintained by tilting the sticks of the hands of the operator to the exact same angle to move the upper wing in the direction of forward or backwards above the lower part of the wing to alter the center of lift.

However, in addition to this shifting the wings back and forward The Wrights included an "elevator" to the rear. It was secured by two wooden rods, which were connected in a straight line to the uprights that joined the wings. When the upper wings were pulled forward to move the kite up from the its front direction, it slid into the air at its highest side and was then pressed downwards and helped turn the wings upwards. This could be familiar to many because the rear elevator operates similar to planes in the present.

The brothers sent letters an email to the Weather Bureau at Washington, in Decemberof 1899, as well Willis Moore, chief of the Bureau and a member of the Weather Bureau, sent them several government bulletins, which included data on wind speeds in various locations. They were looking for more favorable locations and dates to conduct their experiments.

In the month of May, 1900 Wilbur Wright sent an email to Octave Chanute, a resident of Chicago who was the writer of Progress in Flying Machines. Even though Chanute was more well-known within the engineering world the book, which was which was a reprint of his writings that were published between 1891 and 1893, established him as the most authoritative source on the aviation history. Wilbur believed that Chanute was attracted by their exploits and told Chanute in his letter about his plan to test the possibility of a kite that carried people. Wilbur suggested the use of a tower that was high that from where a cable could be connected to the kite for man-carrying. In his letter the control system to be utilized within the kite: the warping of wings to provide control in the lateral direction, and shifting the upper part of the surface backwards and forward to allow longitudinal control. He also requested Chanute whether there was any information available about locations in which winds suitable for carrying out the experiments would be. Chanute in response to Wilbur's request recommended San Diego, California, and St. James City (Pine Island), Florida, to be the most suitable locations for flyingbecause the constant sea breezes. The letter reads: letterthat read "For many years, I've been struggling with the notion that flying is possible to human beings. My condition has increased in severity , and I am afraid that it is likely to cause me to lose a lot of money, even my life. I've tried to organize my life in as to allow me to dedicate my whole time for a couple of months to research this area."

After the brothers quit their bicycle business, they were able to have plenty of time to delve into the subject of equilibrium. Everyday they suggested and discussed the development of new equipment. Orville believed that the shift of the upper surface back in a forward direction over the bottom to achieve a longitudinal equilibrium although it worked in their kites, would not work for a glider carrying a person that would begin and end up in the air. Orville thought that the wings' surfaces should be fixed above each other, and elevators should be located somewhere ahead of the wings instead of in the rear.

After having read that the center of pressure shifts towards the edge of the wings' front when the wings are turned more or less horizontal in flying, Wilbur thought inherent longitudinal stability could be achieved when the front elevator was placed at an angle of negative. With this arrangement of the wings and elevators, as the wings became horizontal in flight and met with the air at a less angle on their lower sides and the elevator would be able to meet with the atmosphere at a higher degree on its top sides.

The Wrights elevator was equipped with three distinct features not found in any other gliders. It was located in the middle of the wings, which made it less vulnerable to be damaged by hitting the ground during take-off or landing. It was also operable rather than fixed like with other models.

Dream big and dream big!

The Spark

In 1903 Wilbur Wright and Orville Wright, two not well-known brothers who were from Dayton, OH, became the first humans to fly with a powered machine, commonly referred to as"the" Wright Flyer. It was not a simple event that happened in a matter of minutes. The Wrights were fascinated by the idea of flying since when they were young. They were curious and curious kids. The boys were highly mechanical invigorated by the efforts of their peers.

The year was 1878 when their dad purchased the toy helicopter for them. The design was created by French aeronautical engineer Alphonse Penaud This toy did not just fall to the ground like one would expect. It was flying in the air until it hit the ceiling. Despite the fact that the toy's fragile structure soon was broken, Wilbur and

Orville decided to make one on their own. A few years later, Orville told the world that it was this toy that ignited an obsession and interest their children to fly.

Through their entire lives they were involved in various projects to earn money. They had already established their own company, the Wright Cycle Company to sell bicycles. They found an extremely profitable business, which also gave them time to work on other ventures. Since the business seemed to be profitable and they began to open an repair shop, and later began manufacturing bicycles.

Although the business was running well but the brothers were in a state of discontent. Their energy was centered by two events that occurred in 1896, the first and most significant of which was the passing of Otto Lilienthal, the celebrated glider scientist who was killed in a plane crash and the other being the success of the launch of unmanned motorized models created of Samuel Langley. The ingenuity of these two brothers was inexhaustible. Fortunately, the bike business helped fund their new passion. In contrast to other people in aviation theirs, the Wright brothers weren't funded by anyone else. They created everything with their own with their own money. The greatest benefit of repairing and producing bicycles only that, but their mechanical abilities were improved. When they built aircrafts, Wilbur as well as Orville often employed the same tools and equipment that they used previously for fixing bicycles. They conducted a lot of their experiments in their back room of their shop , and a majority of the components used in the first airplane that was successful were constructed there.

Methods and inspiration

The Wrights are known for their aviation work began in 1899, the year Wilbur was the first to write The Smithsonian to promote literature. Prior to their time there were many who were not successful in aviation, and the Wrights were unsure if whether they would be successful or not. Wilbur was the leader in the beginning stages of their efforts to resolve the challenges of flight however, Orville was soon recruited as a partner in equal measure. They were attempting to formulate their own theories and over all of the following four years they were devoted to the cause to understand human flight.

The brothers were aware that the ways to improve propulsion and lift required only refinement, but nobody had ever achieved lateral control prior to. In rejecting the idea of inherent stability - the common wisdom they wanted control to be dependent entirely on pilot. Wilbur was a fan to warp the wings that was sparked by his observations of birds, and the involuntary movement of a box to rotate the wings and help stabilize flight. The team tested wing-warping, the precursor of ailerons on a biplane kite that was 5 feet long.

After they had a successful flight in their flying kites, two genius brothers quickly realized that the conditions of the weather in Dayton weren't ideal for flying tests. Therefore, they wrote to their National Weather Bureau in Washington, D.C. asking for the location of the eastern coasts where conditions were stable.

1900

The Wrights were sure of their designs They Wrights created the 17-foot glider, which had an unorthodox forward elevator. Wilbur Wright wrote to the director of the Kitty Hawk weather bureau station inquiring for more information regarding the location, and explaining that he may want to visit the area in the near future to test the feasibility of a man-powered kite. Wright also inquired whether it was possible for him and his friend to find lodging and board near the area until they were able to get set up in an area of camp.

Joseph J. Dosher, responsible for the Kitty Hawk station, replied on August 16 by describing the direction of the current winds. He also detailed the terrain over a long distance.

Dosher provided Wilbur Wright's letter to his neighbor, William J. Tate in the hope that he would could also respond. Tate was probably the most educated person in the area. Tate performed a decent job in writing his letter to Wilbur Wright on the 18th of August. The letter he wrote to Wilbur Wright spoke about the potential of Kitty Hawk. Tate was able to write a convincing letter about the suitability Kitty Hawk region, because of the high winds that prevailed that could be used for the kinds of research Wilbur had mentioned. However, Tate went into detail regarding the sand hills that were treeless as well as the terrain.

These letters written by Dosher as well as Tate led that the Wrights they Kitty Hawk was the place to test their ideas. They instantly decided to travel for a trip to Kitty Hawk, as soon they could construct the glider.

The job at Dayton in getting the parts and the material prepared for the glider was only a couple of weeks. It was decided at that time that Orville would remain in Dayton to take care of Dayton's bicycle workshop until Wilbur was settled in Kitty Hawk before he could take over the Kitty Hawk shop.

Wilbur embarked on his trip in September 1, bringing elements of the glider as well as all the material required to build it, with the exception of some spruce wood which he was hoping to find closer to his goal. As a result, it flew in a kite, focusing on the controls from the ground.

Wilbur's flight time during free flight was just 10 seconds. Then, they returned home feeling defeated, but optimistic about the future as they believed they had successfully achieved lateral and transverse control.

"That Wilbur Wright is in the possession of a force that determines the destiny of the world is undisputed." Major. Gen. Baden Powell, the then president of the Aeronautical Society of Britain

1901

In the year 1901 the brothers honed their focus. In order to solve the issue of lift they increased the camber on the 1901 glider. They also extended the wingspan by 22 feet. which made it the biggest glider ever constructed or designed to fly.

However, at their brand newly constructed Kill Devil Hills camp, the lift was only 1/3 of what was written into Lilienthal data, upon which the wing's design was constructed. The glider then pitched in a wildly erratic manner, and climbed into stalls. Once they returned in the previous camber they regained the longitudinal control and then glided

335 feet. But, the machine was not completely stable. If the pilot lifted the left wing in order to start the anticipated right turn however, the machine tended to slide towards on the left (adverse the yaw). This error made them realize that they were relying on incorrect data, and they reached the point of deciding to quit. But they refused to abandon the cause. They decided to build the wing tunnel in order to collect their own information.

1902

The 1902 aircraft embodies Wrights research. They designed it with efficient 32-foot wings as well as vertical tails to counteract the adverse yaw. The pilot rotated the hip cradle in order to warp the wings. Around 400 glides showed that the design could be used but it was still flawed. There were occasions when the pilot attempted to lift the wing that was lowered to make it out of an angle, but the machine instead moved into the wing's side and was thrown into the ground. To solve the issue, Orville suggested a movable tail to stop this tendency. Wilbur believed that he could link the movement of the tail with the mechanism for warping which allowed the plane to be turned, and then stabilized.

The Wright brothers realized that stability and control were connected, and that a plane could be turned by rolling. Six hundred glides more, that were made in this same period, convinced them with the first airplane that worked. In 1903 , they proved it.

The second that changed everything in the course of history

They were dressed up in jackets and tie on a cold December morning. It was a secret ceremony that changed the course of the world. The swimming pools surrounding the camp were melting and the change in the weather could be their last chance to enjoy the season. The words were difficult to say in the noise of the engine So they exchanged handshakes and Orville set himself up on the flying machine. And then, on a sand beach in 1903 He broke our connection to the earth. He flew. The flight lasted just 12 seconds. The length of the flight was smaller than an airplane.

Overall it was a true masterpiece. It was the first time that an air-powered, manned machine was able to leave the ground under its own force and accelerated forward without losing its speed and came to rest on a level at the same height as the point from which it came.

"Before Wright Brothers, prior to the Wright Brothers, no one was doing anything fundamentally wrong. After when the Wright Brothers, no one has made any fundamental change." -- Darrel Collins US Park Service, Kitty Hawk National Historical Park

Chapter 3: A Failure As Well As The Success Story

It's important to mention that the Wright brothers owe their triumph on Dec. 17th, 1903 in way, in part to numerous mistakes of the aviation pioneers prior to them. In the first place, Otto Lilienthal's tragic demise in a glider crash in 1896 had taught them a lot that they gleaned from the accident that he tried to control his plane using his weight to shift his body. This was not the most efficient solution to the issue.

"Their usage of elevators, the rudder and curved wings surfaces allowed the Wrights to achieve what no other before them could the capability of controlling powered flight." -- Henry Petroski, Aleksandar S. Vesic Professor of Civil Engineering at Duke's Pratt School of Engineering.

The first pioneers of aviation took off out of the ground in self-propelled aircraft however, it was the Wright brothers became the first fly.

"When they took off into flying, they did not know only how to climb off on the ground but also also how to control the airplane during flight." "The evidence of their superiority is the fact that we're still following their model of flight, the Wright brothers. One hundred decades later, they're flying in the same manner as they did and we're not attempting to imitate other claims to prioritization." -- Alex Roland, a Duke historian who is specialized in the technological history.

The most notable thing of their life was the flight over Kitty Hawk's dunes. Kitty Hawk dunes, but before that , they read the pertinent literature, tried their hand with kitesand conducted hundreds of gliders piloted by humans. They were exploring, walking reading, and experimenting constantly. The most notable characteristic in the Wright brothers was the fact that they were way too methodical. They employed the scientific method, and were also excellent mechanists. Contrary to less successful rivals they Wrights were able to blend all the elements needed, including the size of their wings, power, and control.

"As important as the Wrights craft was however, it was not the perfect aircraft. The plane was in fact unstable." The plane was actually unstable. Kenneth C. Hall, the Duke professor and chairman of the Department of Mechanical Engineering and Materials Science who organized the 10th International Symposium on Unsteady Aerodynamics, Aeroacoustics and Aeroelasticity of Turbomachines at Duke in September 2003.

Wright Airplanes

1903 Flyer I 1903 Flyer I was the Wright brothers first powered aircraft. It was the first aircraft with which any person in aviation history was able to sustain a steady, controlled flight. Like the gliders of earlier the model had a variable camber twin canard to regulate pitch, as well as a twin-rudder back to manage the yaw. The roll was controlled by moving the wings.

1904 Flyer II 1904 Flyer II - This is the Wright brothers' second powered aircraft, nearly a complete replica from the Flyer 1, besides a lower camber and more powerful skids. This was not the most great pilot, since the Wrights were aware that they needed a lot to finish before they could have an actual airplane. The Wright was the first to complete a circle before returning to the point from where it started. The Wright completed 105 flights in 1904. On the 20th of

September, they completed the first 360-degree turn in an airplane.

1905 Flyer III The 1905 Flyer III was the Wrights 1905 Flyer III aircraft the third engine-powered machine they built. It was the first aircraft that was practical. The canard as well as the rudder were pushed out of the aircraft in order to enable it to control. In 1908 they modified the Flyer to transport the first air passenger. Orville first flew on the Flyer on June 23rd 1905. Flyer III featured a brand-new frame and an upgraded engine that had slightly larger engines. However, it was the same layout and similar performance and instabilities similar to Flyers I and II. However, it had some flaws that caused a serious crash on July 14, 1905.

1907-1909 Wright Model A 1907-1909 Wright Model A masterpiece showed the world Wright had actually flown. The plane was the first two-seater aircraft and also it was the very first Wright aircraft where the passengers could sit in a straight position.

1909 Military Flyer - This plane was larger than Model A, the Wrights sold the aircraft to the

United States Army Signal Corp to be an early military plane. This was the first airplane that the Wrights created to speed. The moment the US military acquired this aircraft it would be put to use to instruct the first pilots for military use. Test flights with the military aircraft began on June 29, 1909. The Wright brothers Wilbur as well as Orville were on hand, but Orville took over the flying alone. Orville was the sole pilot. US Army agreed to pay the Wright brothers $25,000 for the Wright Military Flyer with a an additional bonus of $5000.

1910-1910 Wright "Transitional" Model A1909-1910 Wright "Transitional" Model A first Wright aircraft constructed with an elevator in the back. It was used for training pilots of the first civilians to fly and was the sole airplane that Wilbur Wright and Orville Wright flew together.

1910-1914 Model B- This was the first aircraft to be mass-produced. It was also notable because it did not have any canard. The elevator was single located in the back, behind an extended twin rudder. Triangular blinkers were positioned on the skid struts in front. The plane was parked on wheels. Similar to all Wright aircraft, this relied on the wing to warp to control the roll.

1911-2012 Wright Model EX - This model was made specifically for display flights. it had a smaller wingspan than other models. This gave the pilots who flew exhibitions more speed. The model had only one seat and made it impossible for the pilot to take passengers. This was the first plane that was flown across continents, traversing America.

1912-2013 Wright Model C - This model replaced Model B. Model B, as the regular Wright Aircraft. It was slightly broader wings, and a higher rudder to improve control of the directional. The blinkers changed into rectangular vanes that were attached to the front of skids. After a string of incidents that followed, it was decided by the US army decided to shut down every "pusher" airplanes which included the entire range of Wright models.

1912 Wright Model D - This plane was light and came with an extremely fast single-seat. Combining their powerful 6-cylinder engine to an airframe that was short-winged and a short-wing airframe, the Wrights reached a rate of about 66 miles an hour. It was however difficult to fly. Only one models was constructed.

1913 Wright Model CH - Even though the Wrights and other pilots had installed floats to the Model B aircraft, this was the first dedicated Wright hydroplane. It was basically the Model C on a single large pontoon. Later models had two slim "stepped" pontoons. This was held up by smaller floating that was mounted beneath the rudder.

It was a powerful lifter for an aircraft powered by water capable of lifting over 800 pounds (363 kilograms) off the ground.

1913 Wright Model E - This model was a single-seat exhibit machine that had a single propeller. Tail booms on the model were positioned to the wings further to accommodate the propeller that was 7 feet long chain-driven. The design was quite straightforward. It was this machine which Orville decided to show his stabilizer's automatic function in the beginning.

1913 Wright Model F - This model was a major shift in the style of Wright Model F, the Wright

Company. They removed blinders and curtains and included the fuselage. They moved the engine to the forward of the wings, and created a semi-standard tail , by placing the rudders on top of the elevator. They also then hinged the elevator. The fuselage was partly coated in aluminium. It was nicknamed "Tin cow."

1913-1914 Wright Model G - This was the only flying craft built in 1913 by The Wright Company. It was created by Grover Loening initially under the direction of Wilbur Wright, and then, after Wilbur's passing, under Orville's direction.

1914-1915 Wright Models H & HS They were like Model F. They were very similar to Model F, besides the fuselage. The model was designed to a point that it curved back towards the tail. The HS was a smaller wing at 32 feet (9.75 meters) for greater speed and a higher rate of climb.

1915 Wright Model K - This was a seaplane built specifically for use by the United States Navy. Model K Model K was Wright Company's first tractor plane with propellers facing towards the forward. The Model K was the very first Wright aircraft with Ailerons.

1916 Wright Model L - The Model L was a single place aircraft, built to meet the U.S. Army's requirement for a lightweight and fast and efficient scouting machine. It was the final aircraft produced from The Wright Company. Orville has already sold his company however, he had a small influences on its design.

1918 Liberty Eagle - Orville Wright and Fred Nash designed this tiny biplane that was unmanned when they worked for the Dayton Wright Airplane Company, which was a manufacturing company set up to build aircrafts in World War I. It was the Liberty Eagle - nicknamed the Bug. It carried 200 pounds (91 kilograms) of explosives, and was controlled by the stabilizing gyroscopic device.

1919 OW.1 Aerial Coupe -This four-place cabin biplane was created by Orville Wright for the Dayton Wright Airplane Company, at the time that World War I ended and the company tried to transition from commercial aviation to military. It never did find its target market, and there was only one constructed. It was the final airplane developed in the hands of one of the Wright brothers.

The Wright Company never made a Model I or a Model J. They were created through the Burgess Company, which licensed Wright patents.

Experiments with flight

"For several years, I've been plagued by the notion that flying is possible for human beings. My illness has grown in severity and I believe that it's going to soon cause me to lose a lot of money, if not even my life. ...." " - Wilbur Wright

When the Wrights reached Kitty Hawk in 1903 for their first ever historic flight they were nearly certain of their success. But, just one year before their 1902 triumph, following failing glider experiments in the years 1900 and 1901 The brothers had almost gave up after defeat. As they returned to Kitty Hawk after their unsuccessful glider experiments in the year 1901, Wilbur could not hide his sadness. He stated: "Not within a thousand years will man fly."

But, after a while they finally discovered the secrets of flying heavier than air. Orville sent a letter to a colleague: "Isn't it astonishing that these secrets have been kept secret for so long to make it possible for us to discover these secrets!"

The Wrights were well conscious of the actual difficulties of navigating the unexplored realm of air however, they did not give up. They Wright Brothers were unwilling to remain in the waiting room and sit and wait for miracles to occur. They did their best.

"If you're looking for absolute security, you'll need to sit in a fence and look at your birds. ..." -- Wilbur Wright

In the year 1900 In 1900, the Wrights constructed the first of an array of fully-sized gliders, to check their wing-warping theories. In this case, they utilized Otto Lilienthal's tables of lift in order to determine the curvature of their wings. They intended to try the glider using kite. However they failed to succeed. Wright Brothers did not succeed this time. While the glider flew well as a kite as well as their warping mechanism worked however, its lifting power did not match the expectations that Lilienthal's tables had made. The glider was able to lift a pilot, but only when the winds were extremely strong however, it was it, however it was extremely dangerous to fly.

The change that they did to Lilienthal's curvature of the wing could have contributed to the 191 Glider that was balanced on the end resulted in the loss that lifted the machine in 1900 The Brothers exactly followed the curvature in the machine of 1901. To increase lift they increased the length of the wings. And the machine was went back in Kitty Hawk in 1901 with their new machine. In the end, the lift was barely increased however the machine was virtually impossible to control in the air.

After this, several weeks went by and the Wrights had to close their camp and returned to Dayton with a sense of despair. They realized that their previous research on aerodynamics was completely incorrect.

After the second time, Wrights decided to make their own tunnel in order to gather information. Wind tunnel testing began in fall of 1901 and finished in 1902. After collecting some data about the design and the behavior of the wings, Wrights had a lot of important data. Thus, at the end of the day, they had the knowledge required to build an effective flying machine.

The Wright Brothers returned to Kitty Hawk in 1902, they brought their parts. They had realized the effects that "aspect ratio" on the wing's ability to lift. Aspect ratio is the relation between the cord of a wing (front-to-rear dimensions) and its length (tip-to-tip measurement). The wings that are long and slender have more lift than shorter slim ones. Vertical stabilizers were an innovative feature on the 1902 model. It was designed to avoid from the "adverse yaw" that was caused by warping the wings.

In the beginning, the 1902 glider had two side-by side vertical stabilizers. However, the Brothers were again faced with a control issue. This issue was referred to by the name of "sidling away into the earth." It was the Wright Brothers also invented the three-axis control systemthat is utilized by all aircrafts operating in the present.

Orville sent a letter to his sibling Kathrine when he left Kitty Hawk in 1902: "We are now the holders of all records! The biggest machine longest distance glide longest duration flying in air lowest slope of descend, and with the strongest wind."

In 1902 after 1902, the Wrights also developed the 1903 Flyer that was a huge 600-lb (274kg) machine that had an wing span at 40.3 feet (12.28 meters). The Flyer was not without issues, but Brother believed that when they were resolved, all that was left to do was include propellers and an engine, and then take off in the air.

Wright Flyer before the 1st Flight Triumph on the sands at Kitty Hawk in December 1903 was the result of three years of scientific inquiry and rigorous testing. Brothers flew four times in the morning on December 17, 1903.

Nearly 66 years after Wright Brothers' first steps into the air, Neil Armstrong and Buz Aldrin made their way to the moon's surface to conclude a new chapter in the epic story.

Power to Fly: The Power to Fly

It is generally accepted that flying is unattainable in the absence of sufficient thrust to sustain the flight speed of the plane. One of the most important factors in determining what the Wright

Flyer could sustain flight is to determine the amount of thrust needed to overcome the aerodynamic resistance referred to as drag. Once the drag is known the required horsepower of the engine is easily established.

In this case, you'll look at the same analysis as what the Wrights used to answer the question of what power was required.

What is drag you ask? It is created from two surfaces of an aircraft when it travels through the air. One is due to the lifting effect that occurs on wings, and the second due to the wind resistance generated by the surface area of the front of the plane.

Drag

The formula that the Wrights employed to determine drag was very identical to that they employed to calculate lift. There is only one difference: the coefficient of lift (CL) has been replaced with the drag coefficient (CD) within the equation. The formula that is used in the base formula is as following:

D = Drag (pounds)

K = the air pressure coefficient

S = wing surface (square feet)

V is the relative velocity of the air that is circling the wings (mph)

CD is the coefficient of drag.

But, for this 1903 Wright Flyer:

K = 0.0033 (Wrights were derived from wind tunnel tests)

S = 512 (wing area of 1903 Flyer)

V = 30.8 (The wind varied between 20 and speeds of 27 mph in Kitty Hawk on December 17 1903. I determined an average wind speed that was 24mph December. 17th, 1903 with a the ground speeds at 6.8 mph. Wilbur was running on the right wing's tip, was able to keep pace with the Flyer when it was moving across the rails that were used for starting to take off.)

On the 23rd of November 1903, Orville wrote Charles Taylor. Orville was able to write Charles Taylor, their employee who constructed the engine in accordance with the blueprints of the Wrights The following was the content of the letter:

"After just a few minutes to make adjustments and also to remove excess oil The engine accelerated the propellers by 351 rev. per minute. With the thrust of 132 lbs. Stocks soared like a rocket and is currently at its highest level in the history of the company. We have allowed for some variance at almost every step of our calculations, which means that as the weight increases we are expecting to reach around 90 pounds. Of course that's being lowered to our most accurate estimates."

Power

Power is what it says? Power is force multiplied by speed. The force needed for overcoming drag could be calculated by multiplying drag total by velocity

If you convert this number to horsepower it is 7.3. The engine of 1903's Wright Flyer had about 12 horsepower. It could reach 16 hp when it was first started, but then drop down to 12 horsepower after only a few minutes. Thus, Wrights knew it would be a tough call. In November, Orville left a note to his sister and father:

"Mr. Chanute says that nobody else has attempted to construct an airplane with such narrow margins like we have according to these calculations." (Octave Chanute, who was also a good friend and aviation historian as well as experimenter.)

The 1904 Flyer was different in many ways to Kitty Hawk Flyer but they did upgrade the engine, so that it was able to produce 15-16 horsepower. Wilbur addressed a note to Chanute on August 8th, 1904:

"We have encountered a lot of difficulty in achieving sufficient speed to make real start. Although the new machine is able to lift at speeds of around 23 miles, it's only after it reaches the 27-28 mile mark that resistance drops to below that of that of the thrust."

The brothers came up with the answer by using the catapult launcher system for giving the Flyer an extra boost when it takes off.

With all the information and skills I have gained from thousands of flights over the last 10 years, I wouldn't imagine today making my first flight with an unorthodox machine in a wind of 27 miles even if I knew it had been piloted and was safe. - Orville Wright

First practical fixed-wing aircraft

The Wrights began rebuilding their aircraft in May 1905. They replaced the propellers, engine and the hardware. The Flyer III Flyer III was designed to be able to overcome the difficulties they encountered in 1904. The elevator as well as the rudder were bigger, which gave the plane the ability to have more "authority" in its pitch as well as yaw controls. The brothers also encountered certain issues the issue of "side slips" in the sense that the plane was known to slide sideways in the course of a turn. To fix the issue they put semi-circular "blinkers" between the elevator's surfaces, to keep the plane moving forward. They also noticed that propellers tend to flatten and twist when they were spinning, which reduced thrust. To prevent this problem, they attached tabs, referred to as "little jokers" to the edges that trailed on the propellers.

The Wrights tried to fly again in June, however the Flyer wasn't working. They flew 8 times with the brand new Flyer 3, the longest under 20 seconds and each flight ended in the aircraft being damaged. On the 14th of July, Orville smashed into the ground at a rate of more than thirty miles an hour. It smashed into on the elevator in front. In the Flyer III bounced three times before slamming Orville out of the top side. Orville was discovered by his brother in the dirt, confused and confused.

Following this devastating failure The brothers repaired the Flyer once more, increasing the elevator once more , and expanding it from 7-1/2 inches to a distance of nearly 12 feet of the wings. They also widened the rudder and created the first "bent-end" propellers too.

Orville and Wilbur began flying again in August. It was clear that the improved Flyer 3 was truly airworthy. In just a few days the pair were flying several circuits over the prairie, and landing without serious accidents. On the 26th of

September, Wilbur flew for 18 minutes and ran the tank to the end as the very first time they've done this in aviation history. Orville reached the half-hour mark on the 3rd of October.

It is believed that the desire for flying was an idea passed down to us by our ancestral ancestors whom... gazed enviously at the birds flying in the sky... along the endless highway of air. -- Wilbur Wright

The Wright brothers began sending invitations to a variety of people to invite them to be awed by their achievements. The word began to spread all over the city, claiming that an extraordinary feat was executed through The Wight brothers. They issued around 30 invitations to watch the flying on the 4th of October However, it was discovered how the amount of people who attended could reach several hundred.

On the 5th of October, a tiny crowds of people included Torrence Huffman, and Dave Beard, came to observe Wilbur as well as Orville fly. The pilot's first flight in the morning was brief only 40 seconds. The Flyer III rose gently into the air, then made an 180-degree turn, then returned to land safely. However, in later in the day, Wilbur was able to fly 30 times while on the ground for almost 39 mins. He flew over 24 miles and only landed when he ran empty of fuel. It would be the longest air flight in history of aviation.

In 1905, the Wright Flyer III was the remarkable result of careful intricate engineering. It was built in small pieces, beginning with the experiments on kites in 1899 It was the first aircraft to be capable of having its wings lifted and floating in the air with its own power, ascending, descending and changing direction at any time under the direction of a pilot. It also had the capability of landing without any crash. The Flyer 3 was the world's first airplane that was actually used.

These airplanes, built in 1904/1905, were the first truly functional flying machines that could be operated at will. The brothers needed to be extremely dedicated on the airplanes in order to make them work according to the way they needed to function. It was a lot of hauling and pushing conditions which, in some ways are more difficult than those at Kitty Hawk. The idea of climbing up and making around in a circle, and flying for a couple of hours at a time is a fantastic moment. --James Tobin, author.

Chapter 4: The Last Years

Wilbur

There was no one of them were married. Wilbur was once quoted as saying that he could not find the enough time to have wife and an aircraft. After a short flight for training within Berlin, Wilbur Wright gave an opportunity to an German pilot in the month of June, 1911. Wilbur never flew again following the flight. He was soon occupied with issues of business with Wright's company Wright Company, dealing with numerous lawsuits. In the course of settling patent litigation, which placed a immense pressure on the brothers Wilbur made a note in one letter letter to an French friend "When we consider about what we could have done in the event that we were able to dedicate this time to research, we are very sad, yet it's always simpler to handle things than with humans who can't control his life in the way the way he'd like to."

Wilbur was a busy man the year prior to his death travelling and spending six months in Europe taking care of various legal and business issues.

Additionally, he spent lots of times at New York, Washington and Dayton. The business-related issues stress were really affecting Wilbur physically. Orville even said that Wilbur often returned home looking pale. Since the brothers made a lot of revenue from their business ventures The family then decided to construct a bigger home. When Wilbur lived I Europe, their house was constructed on Oakwood, Ohio. Dayton neighborhood located in Oakwood, Ohio. But, Wilbur would not live to see the building completed in 1914.

He was ill during an official excursion from Boston during April of 1912. The cause of the illness was eating bad oysters at a dinner. When he returned to Dayton in the early part of May 1912, tired in body and mind the man felt sick again and consulted a doctor. He was diagnosed with typhoid fever.

Wilbur Wright was killed at the age of 45 in the Wright family home on May 30. The Wright's father Milton wrote regarding Wilbur within his journal:

"A very short time in life, filled with consequences. A unwavering intellect, a steadfast willpower, great self-confidence as well an impressive humility, knowing the right path and following it with determination He lived and died."

Orville

Orville was elected president in The Wright Company upon Wilbur's death. Orville was not so enthralled in running the company, and didn't have the same capabilities as his deceased brother Wilbur was, and he eventually ended the business in 1915. After 42 years of living in their home at 7 Hawthorn Street, Orville, Katharine and their father Milton relocated into Hawthorn Hill in the spring of 1914. Milton passed away in his sleep on the 3rd of April 1917, at the age of 88. However, up until his death Milton was active by writing pieces for religious publications, and taking walks in the morning. He was even a part of the Dayton Woman's Suffrage Parade, together with Orville as well as Katharine. Orville completed his final flight as pilot in 1918, in an aircraft built in 1911. Model B.

He resigned from his business and was a senior statesman of aviation. He served on various committees and boards, such as the National Advisory Committee for Aeronautics (NACA) which was the agency that served as the precursor that is now the National Aeronautics and Space Administration (NASA) and the Aeronautical Chamber of Commerce (ACCA) which was the predecessor of Aerospace Industries Association (AIA). Aerospace Industries Association (AIA).

Katharine was the younger sibling of Orville, was married Henry Haskell of Kansas City Henry Haskell, an ex- Oberlin classmate in 1926. Orville was furious over this, since he wanted Katherine was right beside him throughout the day. He was even unable to attend the wedding, nor communicate with Katherine. He finally accepted to visit her, at her husband's request, right before her death from pneumonia on March 3rd 1929.

Orville Wright was a member of NACA during the period of 28 years. In 1930, he was awarded his first Daniel Guggenheim Medal established in 1928 by the Daniel Guggenheim Fund for the

Promotion of Aeronautics. The year after, he appointed to the National Academy of Sciences.

On the 19th of April 1944 Second production Lockheed Constellation, piloted by Howard Hughes and TWA president Jack Frye, flew from Burbank, California, to Washington, D.C. in 6 hours and 57 mins. On the return journey the aircraft made a stop in Wright Field, in order to offer Orville Wright his final airplane flight.

Orville was deeply sad about the devastating effects on the world during World War 2 and ones during an interview

"We were a bit naive to believe that we could have invented something that would bring long-lasting peace to earth. However, we were not. ....No I'm not regretful regarding my role in the creation of the plane, but there is no one who would deplore more than I do at the damage it caused. I am a bit similar to how I do with fire. This is because I am sorry for the great damage done by fire, however I believe it's good for us all that somebody has discovered how to ignite fires, and we've discovered how to use fire to a myriad of crucial applications."

The last major project of Orville was to oversee the preservation and reclamation of 1905's Wright Flyer III, which is regarded as the first real-world airplane.

Orville died on the 30th of January 1948, more than 35 years following his younger brother. Both brothers are interred in the family graveyard at Woodland Cemetery, Dayton, Ohio.

Here are two mechanics for bicycles in the Midwest. They don't have any money aside from what they come up with. They do not have formal training even though both, Wilbur especially, are highly read, and as reading as any scientific researcher has ever been. Just because of their power, they easily beat every one of the world's top scientists. Hyram Maxim Samuel Langley, Thomas Edison, and all the other scientists who attempted to solve the issue of flight before, the Wrights brothers are able to sail straight through the challenge. This is exactly what Americans are looking to be told about their own. Little guys can achieve it. What you require is confidence in yourself and to keep your mind on you and outthink the other person. If you're not rich and you don't have power in the political arena; and even if you do not have all the resources the resources, what you do haveis your wits, your Yankee imagination. We use this repeatedly to show our children that they can live with what they've got, and can accomplish amazing things even without fortune or power. --Nick Engler Wright Brothers Aeroplane Company

Aviation Business

The Wright brothers, who were in the aviation business after their flights that were successful, began receiving numerous requests. One offer was made by two brothers from Detroit important stockholders of the Packard Automobile Company, Russell A. and Frederick M. Alger. The brothers, shortly after they made their proposition were the first people in the United States to order a Wright machine for personal use. However, it is important to point out that the first American firm to build the aircraft was founded through Clinton R. Peterkin. Peterkin was just twenty-four years old and appeared like a teen. In the space of a year or two prior to that, he worked for J. P. Morgan & Company as an "office youngster". He was extremely hard-working and was very intelligent.

In those days, the Wright brothers were very well-known and this young man was eager to meet the Wright brothers. He was fortunate enough to learn the fact that Wilbur Wright was staying for some time in the Park Avenue Hotel in New York and, in the month of the month of October 1909, he made a trip to meet the Wright brothers. But, Wilbur and Orville did not wish to form any kind of company, and Wilbur didn't think that he could take Peterkin seriously. But, at the end of the meeting Wilbur said he was able to take a look and see what it was he could do.

It was evident that Peterkin was quite serious, when in a very short time, an impressive list of moneyed men were enrolled as subscribers in the proposed flying-machine company. This list even included Cornelius Vanderbilt, August Belmont, Howard Gould, Theodore P. Shonts, Allan A. Ryan, Morton F. Plant, Andrew Freedman and E. J. Berwind. Shonts was president of the New York Interborough subway. Ryan, a son of Thomas F.

Ryan, was a director of Bethlehem Steel Corporation. Plant was chairman of the Board of Directors of the Southern Express Company, and Vice President of the Chicago, Indianapolis & Louisville Railroad. Berwind, as President of Berwind-White Coal Mining Company, accumulated a great fortune from coal contracts with big steamship lines. Freedman made his money originally as a sports promoter and then, in various financial operations. As you see all the moneyed men were very serious and famous people.

In addition to those people, the Wright brothers wanted to see also their friends Robert J. Collier, publisher of Collier's weekly and the two Alger brothers of Detroit in their company. So, their names were also added.

On November 22, 1909, only about a month after Peterkin's first talk with Wilbur, The Wright Co. was incorporated. The capital stock represented a paid-in value of $200,000. The company opened impressive offices in the Night and Day Bank Building, 527 Fifth Avenue, New York, but the factory would be in Dayton.

In January, 1910, Frank Russell, Alger's cousin, who was appointed as a factory manager, arrived in Dayton and went to see the Wrights at their office over the old bicycle shop. But there was no space for him to work in, so the brothers suggested a room at the rear of a plumbing shop down the street where he might make temporary headquarters. According to Frank Russell, Wilbur Wright came there a day to two later carrying a basket filled with letters, directed to The Wright Co. Russel told that Wilbur told him the following:

"I don't know what you'll want to do about these, maybe they should be opened. But of course if you open a letter, there's the danger that you may decide to answer it, and then you're apt to find yourself involved in a long correspondence."

For the first period, The Wright Co., rented floor space in a factory building, but almost immediately the company started to build a modern factory of its own, and it was ready for use by November 1910.

Soon, the brothers decided to give public exhibitions. So, they got in touch with Roy Knabenshue, a young man from Toledo, who had been making balloon flights since his teenage years. The result of their conversation was that Roy took charge of the work of arranging for public flights. The field for flights was selected Montgomery Alabama, nowadays known as Maxwell Field.

Orville Wright was the one to train pilots for exhibition flights. The first pilot Orville trained was Walter Brookins of Dayton. However, Orville Wright also was making frequent flights until 1915, personally testing every new device used on a Wright plane.

There were flights also in Huffman field. Three flights at Huffman field in May, 1910, were especially noteworthy. A short one by Wilbur – one minute twenty-nine seconds – on May 21, was the first he made alone since his sensational feats starting from Governors Island. And it was his was the last flight as a pilot Wilbur ever made.

But on May 25 he and Orville flew for a short time together, only one occasion when two brothers were in the air at the same time, together.

The average charge by The Wright Co. for a series of exhibition flights at a county fair or elsewhere was about $5,000 for each plane used. In 1910, Dayton people saw the first flight over the city itself. Thousands of people witnessed it for the first time.

In late October, 1910 at Belmont Park, New York, Wright planes participated in a great International Aviation Tournament. All other planes taking part were licensed by The Wright Co.

Orville Wright now devoted his time mainly to supervision of engineering at the factory of The Wright Co. Wilbur was kept busy looking after the prosecution of suits against patent infringers and March, 1911, he went to Europe in connection with suits brought by the Wright Company of France. From France he went to Germany.

In 1911, The Wright Co. benefited from another aviation record. Cal P. Rodgers, who received some of his flying training at the Wright School, made – between September 17 and November 5 – the first transcontinental airplane trip, from New York to California. That was really something for everybody. The retail price of a plane in Wright Company was $5,000. So, now the Wrights were wealthy enough to look forward to their retirement.

Then, black period started, Wilbur became ill, and on Thursday early morning, May, 30, 1912, he died. He was aged only forty-five years and forty-four days. Messages of condolences were coming from all over the world. In his will, Wilbur made bequests of $50,000 to each of his two older brothers and to his sister; and $1,000 to his father "to use for little unusual expenditures as might add to his comfort and pleasure." And the entire residue he left to Orville. Orville Wright succeeded his brother as president of The Wright Company.

In 1913, business affairs were really complicated, because that year Dayton had the worst flood in its history. The Wright factory was not overflowed, but not many of the employees could reach the building.

In 1914, Orville Wright bought the stock of all other shareholders in The Wright Co., except that of his friend Robert J. Collier. Orville wanted to get entirely out of the business.

In 1915, Orville received an offer and gave an option to a small group of eastern capitalists that included William Boyce Thompson and Frank Manville, the latter president of the Johns-Manville Co. The deal was closed in October 1915. However, after Orville gave his option to the eastern syndicate, Robert J. Collier came to tell him an important piece of news, and to urge him not to sell the company.

Collier after talking to his friend, the wealthy Harry Payne Whitney, urged him to buy the stock of The Wright Co., thus gaining ownership of the Wright patents, and then immediately make the patents free to anyone in the United States, who wished to manufacturer airplanes. However, Orville told Collier that he was a bit late. The option was already legally drawn and the holders presumably wished to exercise it.

It's not a terribly attractive period in Orville's life. He looks to be sort of overly stubborn and small-minded at certain points. And yet, it's probably true that if he hadn't behaved the way he did, at least for some period until historians looked at the record more closely, the Wrights might not have gotten the credit they deserved. —James Tobin, author

Wright Brothers' struggles

Flight problems

The Wrights followed Sir George Cayley's lead, they initially reduced the obstacles to flight to three broad categories:

- Set of lifting surfaces, or wings
- Means of propulsion
- Method of balancing and controlling the aircraft

Their earliest experimenters focused only on one or another of these problems and did not consider the final design from the outset. They did not pay attention to tis there problems equally and simultaneously. However, they then, recognized that each of these areas had to be successfully addressed to build a working airplane. The brothers thought that aerodynamic and propulsion problems couldbe comparatively easier to solve, so they first concentrated on how to maintain balance and control.

The patent

During their so many experiments in 1902, the Wright brothers managed to succeed in controlling their glider in all three axes of flight: pitch, roll and yaw. Their most amazing discovery was the simultaneous use of roll control and yaw control. A forward elevator controlled pitch.

In March 1903, they applied for a patent on their method of control. The application, which they wrote themselves, was, however, rejected. In early 1904, they hired Ohio patent attorney Henry Toulmin, and on May 22, 1906, they were granted U.S. Patent 821,393for a "Flying Machine".

The patent's importance lies in its claim of a new and useful method of controlling a flying machine, whether powered or not. The patent states that other methods instead of wing-warping could be used for adjusting the outer portions of a machine's wings to different angles on the right and left sides to achieve lateral roll control.

The concept of lateral control was initially very important to all aircraft designs; without it they could not be easily or safely controlled in flight.

Patent struggles

The patent struggles were really severe. They took much time. In 1908, the Wrights warned Glenn Curtiss not to infringe their patent by profiting from flying or selling aircraft that used ailerons. Curtiss refused to pay license fees to the Wrights and sold an airplane to the Aeronautic Society of New York in 1909. The Wrights filed a lawsuit, beginning a years-long legal conflict. The brothers sued foreign aviators, who flew at U.S. exhibitions, including the leading French aviator Louis Paulhan.

The brothers' licensed European companies, which owned foreign patents the Wrights had received, sued manufacturers in their countries. The European lawsuits turned out to be only partly successful. A German court ruled the patent not valid due to prior disclosure in speeches by Wilbur Wright in 1901 and Octave Chanute in 1903. In the U.S., the Wrights made an agreement with the Aero Club of America to license airshows. Promoters of approved shows paid fees to the Wrights. The Wright brothers won their initial case against Curtiss in February 1913, but the decision was appealed.

In January 1914, a U.S. Circuit Court of Appeals upheld the verdict in favour of the Wrights against the Curtiss company, which continued to avoid penalties through legal tactics.

They wanted fame in a lot of ways. They wanted recognition, but I think they wanted it on their terms, and they wanted it in their manner, in their way. —Leonard Bruno, author

Aftermath

All these lawsuits damaged the public image of the Wright brothers, who previously were generally regarded as heroes. Critics said the brothers actions may have retarded the development of aviation, and compared their actions unfavourably to European inventors, who worked more openly.

The Curtiss and Wright organisations merged in 1929 to form the Curtiss-Wright Corporation, which exists even nowadays.

And this meant, anyone who was flying for profit, making exhibition flights, or making airplanes that were capable of using the Wrights' airplane control system which they had patented, they would have to pay royalties to the Wrights. — Fred Howard, author

5 life lessons we can learn from the Wright brothers

Exactly 112 years ago Wilbur and Orville Wright made the first controlled, powered human flight. Of course, as everything in this modern world, this claim is disputed and it is possible that someone, somewhere else, may have beaten them to it. However, the Wrights brothers are But the Wright are known for making a first powered and controlled plane. These two men were amazing thinkers; they were able to manage their business. There are many important treats that one can learn from them. So, what important life-lessons we can earn from the Wright brothers?

1. Wrights were true pioneers

Pioneers are generally people that are the first to use, to apply, to explore something. The true pioneers are the ones that dare. Wilbur and Orville Wrights, indeed, dared. They were the first to achieve so much success in aviation.

The true pioneers push forward. Despite the efforts of others, the true pioneers stay focussed on their goals. If you want to be a true pioneer, you should embrace you failures and learn from them, as the Wright brothers did after every single failure experienced.

2. Defying convention

In 1903 convention was certainly not to fly. Human being was not born to fly, that is the only reason why he so desperately wants to. Those that dared to defy this human flight convention and to pioneer its advancement, knew that an increase in engine power was more important than the development of a reliable system of pilot control.

Defying a convention is not just about societal convention, as taking and following wise advices is a very crucial process in every human's life. But defining you own path and then, being able to even follow it till the end is much more important, while not without its own risks, is part of the integrity of the journey that you define.

3. Overnight success

Achieving you target, defying or pioneering in something can be a life changing experience. These brothers were really amazing if we take into consideration how huge success they had. Their seemingly overnight success was not however overnight. They spent many years to plan, to explore, to try, then fail, then again explore, try and succeed. They poured their hearts and souls into their work.

Nowadays, the society reduced such important treats like patience, waiting a bit longer for something to be accomplished. Modern people increased the demands for instant gratification and reduced attention spans. Overnight successes are unusual and there are exceedingly few that occur without huge amounts of hard graft. It is evident that overnight successes normally take years of work and dedication.

4. It is never over

112 years ago, the Wright Brothers made history. But they did not rest on their success. There was so much more to achieve and develop. Sure, that day when human flight became reality they must have celebrated, but they went back to work again, to make many efforts again, and to achieve more. Just imagine, Orville Wright was born into the horse-and-buggy era and died at the dawn of supersonic flight. You can be measured in contrast with the last thing you did. Your journey in life will continue no matter how many successes and failures you have. It is never over.

5. Be mad

As Albert Einstein once said, "For an idea that first does not seem insane, there is no hope".

The Wright Brothers had a crazy, insane, mad idea. They had a dream to fly, which was unbelievable insane, but they succeeded. They believed, they tried and managed. They were mad, passionate about their idea. Be prepared to be mad; it is perfect madness after all.

If we all worked on the assumption that what is accepted as true is really true, there would be little hope of advance. - Orville Wright

The Wright Brothers Timeline

1867

April 16

Wilbur Wright was born near Milville, Indiana.

1869

Spring

Wright family moved to Dayton, Ohio.

1871

Bishop Milton Wright sits on the porch of 7 Hawthorne Street, where the family lived from 1871 to 1914.

August 19

Orville Wright was born.

1874

August 19

Wilbur and Orville's sister, Katharine Wright was born.

1878

Upon his return from a church business trip, Bishop Milton Wright brought home a toy Penaud helicopter. The toy inspires Wilbur and Orville's first interest in flight.

1881

June

Wright family moved to Richmond, Indiana, where Orville took up kite-building.

1884

June

Wright family returned to Dayton.

1885

Wilbur, Orville and Katharine attended Central High School, located at Fourth and Wilkinson Streets in downtown Dayton. While neither Wilbur nor Orville completed their high school education, Katharine graduated from Central in 1892.

September

Wilbur took special "postgraduate" courses at Dayton Central High School and studies Greek and trigonometry.

1886

With his friend Ed Sines, Orville started The Midget, a school newspaper, with a press given to him by his brothers and type from his father.

1887

September

Orville started Dayton Central High School.

1889

March 1

Orville began to publish the weekly West Side News. Editor and publisher, he maintained an active interest in printing and publishing for several years.

July 4

Wilbur and Orville's mother, Susan Catherine Koerner Wright, died at age 58.

1890

April 30

Orville and Wilbur turned West Side News into an evening newspaper, The Evening Item, although publication ceased in August.

September

Orville began final year of high school as a special student in Latin, however, left school before graduation.

December 13

Paul Laurence Dunbar started the Dayton Tattler, printed by his classmate, Orville Wright.

1892

December

Orville and Wilbur opened a bicycle shop, the Wright Cycle Company. They remained in the bicycle manufacturing and repair business until 1907. The business gave them the funds necessary to carry out their early aeronautical experiments.

1893

Wilbur and Orville attended World's Columbian Exposition in Chicago.

1894

October 20

Wilbur and Orville started a weekly magazine, Snap Shots.

1895

Orville invented a calculating machine that multiplies and adds.

1896

Wright brothers began to manufacture their own brand of bicycles—first the Van Cleve and the

"Wright Special," and later the less expensive St. Clair.

August–October

Orville got seriously sick with typhoid fever.

August 10

Otto Lilienthal, German engineer and aeronautical pioneer, died from injuries suffered in a crash while testing his latest single-surface glider. The tragedy renewed the Wright brothers' interest in Lilienthal and the problem of human flight.

1897–1898

While running their bicycle business, Wilbur and Orville studied the problems of mechanical and human flight. After reading extensively and studying bird flight and Lilienthal's work, the brothers were convinced that human flight was possible and decide to conduct some experiments of their own.

1899

May 30

Wilbur wrote Smithsonian Institution inquiring about publications on aeronautical subjects.

July–August

Brothers built, and then Wilbur flew a biplane kite in order to test the "wing-warping" method of controlling a flying machine. This experiment encouraged the Wrights greatly. They decided to proceed with constructing a flying machine with a pilot.

November 27

Brothers wrote the U.S. Weather Bureau for information on an appropriate place to conduct flying experiments.

1900

May 13

Wilbur wrote to Octave Chanute, a civil engineer and aeronautical pioneer. They estimated an important friendship, lasting until Chanute's death in 1910.

September 6

Wilbur left for Kitty Hawk, North Carolina. Orville arrived later, and they stay with William J. Tate until their camp was ready in early October.

October

Wrights began their experiments, flying their glider as a kite and as a man-carrying glider. They stayed until October 23.

1901

June 26

Octave Chanute met the Wrights for the first time in Dayton.

July

Wilbur's articles, "Angle of Incidence," published in Aeronautical Journal, and "Die Wagerechte Lage Wahrend des Gleitfluges," published in Ilustrierte Aeronautische Mitteilungen, were the brothers' first published aeronautical writings.

July 10

Wrights arrived in Kitty Hawk and began experiments with a larger glider. From fifty to one hundred flights were made in July and August, ranging in distance from twenty to almost four hundred feet.

August 4–11

Chanute visited the Wrights at Kill Devil Hill and witnessed some of their glider experiments.

August 20

Wrights left Kitty Hawk.

September 18

Wilbur addressed the Western Society of Engineers on the brothers' 1900–01 gliding experiments.

October–December

Wrights conducted tests on airfoils and built a wind tunnel.

1902

August 28

Wilbur and Orville arrived at their Kitty Hawk camp.

September 8–19

Wrights assembled their new glider.

September 19–October 24

Wright brothers made between seven hundred and one thousand glides, increasing their distance to 622-1/2 feet.

October 28

Wrights left Kitty Hawk.

December

Wrights conducted experiments with propellers and began to build their 1903 four-cylinder engine.

1903

March 23

Wright brothers applied for a patent on their flying machine (patent issued on May 22, 1906).

September 25

Wilbur and Orville arrived at Kitty Hawk.

September 28–November 12

Wrights experiment edwith 1902 glider.

October 9–November 4

Brothers assembled the 1903 machine and installed the engine.

November 5–December 9

Propeller shafts broke twice. The brothers returned to Dayton to repair them and obtain replacements.

December 14

Wilbur made the first attempt to fly a powered machine from slope of Big Kill Devil Hill. But the attempted turned out to be a failure. Machine stalled after 3-1/2 seconds in the air and landed 105 feet below.

December 17

Wilbur and Orville made the first free, controlled and sustained flights in a power-driven, heavier-than-air machine. Three men from the Kill Devil Life Saving Station and two from Nags Head witnessed the four trial flights. First trial was made by Orville at 10:35 A.M., staying twelve seconds in the air, it flied 120 feet. John T. Daniels photographed the first flight with Orville's camera. Wilbur made the longest flight in the fourth trial, fifty-nine seconds in the air and 852 feet.

December 21

Wrights left Kitty Hawk.

1904

January 22

Wrights employed Harry A. Toulmin, a patent attorney, to work on their patent case.

March

Wrights applied for French and German patents on their airplane.

April–May

At Huffman Prairie, a large meadow near Dayton, Wilbur and Orville built a new heavier and stronger machine with a more powerful motor.

May–December

Wrights made practice flights with their new 1904 machine at Huffman Prairie—total flying time is forty-nine minutes. Wilbur made the first turn in the air on September 15 and the first complete circle on September 20. Longest flight of the year was five minutes four seconds, 2-3/4 miles—almost four circles around the field.

1905

January

U.S. Board of Ordnance and Fortification rejected the Wrights' offer of sale of their airplane.

October 5

Wilbur made the longest flight of the year: 24-1/5 miles in 39 minutes, 23-4/5 seconds, more than twenty-nine times around the field, at an average speed of thirty-eight miles per hour.

October 27

U.S. Board of Ordnance and Fortification declined the Wrights' second offer of their airplane.

1906

January 6

Wrights joined the Aero Club of America.

May 22

U.S. Patent Office granted the Wrights patent, No. 821,393, for a flying machine.

1907

Brothers travelled to Europe to negotiate for the sale of the Wright airplane abroad. Hart O. Berg and Flint & Company are their agents.

November–December

Wilbur met with officials from U.S. Signal Corps and Board of Ordnance to discuss their airplane's capabilities.

December 23

U.S. Signal Corps advertised for bids for a military heavier-than-air flying machine to be submitted by February 1.

1908

January 27

Wrights submitted their bid to U.S. Signal Corps to supply a heavier-than-air flying machine, weighing between 1,100 and 1,250 pounds. It could carry two passengers and fly at a speed of forty miles per hour.

April 9

Wilbur and Orville arrived in Kitty Hawk to brush up on their flying skills.

May 14

Wrights carried a passenger on a flight for the first time: Charles W. Furnas flied with Wilbur.

May 29

Wilbur arrived in Paris to demonstrate the capabilities of the Wright airplane in Europe.

August 8

Wilbur made his first flight at Le Mans, France—the Wrights' first flight in Europe.

August 27

Wright 1908 airplane was assembled and ready for testing at Fort Meyer, Virginia.

September

Orville made U.S. Army test flights at Fort Meyer and established records with and without passengers.

September 17

Orville was seriously injured and his passenger, Lt. Thomas Selfridge, was killed in an airplane crash at Fort Meyer.

November 1

Orville and sister Katharine arrived in Dayton after his discharge from the hospital in Fort Meyer.

November 30

La Compagnie Générale de Navigation Aérienne, the French Wright company was organized.

December 31

Wilbur won 1908 Michelin Cup and a prize of twenty thousand francs with his flight of 123 kilometers, two hundred meters in two hours, 18 minutes, 33-3/5 seconds. He extended this same flight to break a new world record in a time of two hours, 20 minutes, 23-1/5 seconds over 124 kilometers, 700 meters.

1909

January 12

Orville and Katharine joined Wilbur in Paris.

January 14

Wilbur arrived at Pau, France. Orville and Katharine joined him a few days later.

February–March

Wilbur made a series of training flights with three French student pilots at Pau.

March 4

Congressional Medal was awarded to the Wrights by resolution of Congress (H.J. Resolution 246), "in recognition of the great service of Orville and Wilbur Wright, of Ohio, rendered the science of aerial navigation in the invention of the Wright aeroplane, and for their ability, courage, and success in navigating the air." Medal was presented to the brothers on June 18.

April 1

Wilbur arrived in Rome to make demonstration flights and train two Italian pilots. Orville and Katharine arrived on April 9.

May 11

Wrights arrived in New York.

May 13

Flugmaschine Wright Gesellschaft, the German Wright company in Berlin was formed.

June

Wrights performed propeller tests in Dayton to determine the main cause of the Fort Meyer accident in order to prevent similar future accidents.

June 17–18

Two-day celebration thrown by the city of Dayton to honour the Wright brothers

June 20

Wilbur and Orville arrived in Washington, D.C. to resume trial flights at Fort Meyer for U.S. government.

June 26

Glenn H. Curtiss sold his Curtiss airplane, the first commercial sale of an airplane in the United States, to Aeronautic Society of New York for $7,500. Sale set in motion the beginning of the Wrights' patent suit against Curtiss.

July 27

With Lt. Frank P. Lahm as his passenger, Orville flew for one hour, 12 minutes, 37-4/5 seconds. Flight fulfilled the Army's requirements and was witnessed by President Taft, his cabinet, and other public officials.

August 8

Orville and Katharine left for Europe for demonstration flights and sold negotiations in Germany.

August 18

Wrights began a patent suit against Herring-Curtiss Company and Glenn H. Curtiss, by filing a bill of complaint to prevent them from manufacturing, selling, or using in exhibition the Curtiss airplane.

August 19

Wrights filed suit against Aeronautic Society of New York to prevent further exhibition and use of the Curtiss airplane owned by the society, because it infringed on Wright patents.

October 4

As part of the Hudson-Fulton Celebration, Wilbur flew round-trip demonstration flights from Governors Island, New York, to the Statue of Liberty and Grant's Tomb, New York City. More than one million spectators were present.

October 8–November 2

At College Park, Maryland, Wilbur trained first U.S. Army fliers.

November–December

Wright Company moved forward on patent lawsuits. Wilbur and Orville gave affidavits and attended trial for The Wright Company v. Herring-Curtiss Company and Glenn H. Curtiss patent suit.

1910

January

Wright Company and Wright brothers continued their involvement in patent suits.

March

Wright Exhibition Company formed, with Roy Knabenshue as manager.

March 26–May 5

Orville conducted flight training school in Montgomery, Alabama, for pilots who would fly for Wright Exhibition Company.

1911

March 12

A Wright B flyer flew over Huffman Prairie Flying Field.

1912

January–April

Wilbur and Orville testified for patent infringement lawsuits.

May 29

Wilbur dies of typhoid fever in Dayton.

June 1

Wright's funeral procession passed through the gates of Dayton's Woodland Cemetery.

1913

February 10

Orville and Katharine left for Europe on business and returned on March 19.

March 25–27

Miami River flood and caused considerable damage to the Wright family home and property in Dayton.

1914

January 13

U.S. Circuit Court of Appeals of New York ruled in favor of the Wright Company in its suit against Herring-Curtiss Company and Glenn H. Curtiss.

April

Orville, Katharine and the Bishop moved into Hawthorn Hill located in Oakwood, a suburb of Dayton.

November 16

Wright Company filed a complaint against Curtiss Aeroplane Company for continuing to manufacture, use, and sell flying machines which infringe on Wright patent.

1915

April–May

Orville involved in patent infringement lawsuits.

August

In its 1914 annual report, the Smithsonian Institution stated that Samuel P. Langley's aerodrome was "the first aeroplane capable of sustained free flight with a man." This statement led to the controversy between Orville and the Smithsonian Institution that was not resolved until 1942.

October 15

Orville sold his interest in the Wright Company, but served as consulting engineer.

1916

August 7

Wright Company merged with Glenn L. Martin Company, becoming Wright-Martin Aircraft Corporation. Orville served as chief consultant engineer.

1917

April 3

Bishop Milton Wright died in Dayton.

1920

January 13

Orville gave depositions for patent lawsuits.

January 29

President Wilson appointed Orville a member of National Advisory Committee for Aeronautics. He served until his death in 1948.

May 23

Wilbur and Orville's brother Reuchlin Wright died in Kansas City, Missouri.

1921

February 2

Orville gave depositions for patent lawsuits.

1925

January 20

Orville issued a patent for a mechanical toy. The toy was produced and sold by the Miami Specialty Wood Company in Dayton, of which Lorin Wright was the president.

May

Orville and the secretary of the Smithsonian Institution publicly disagreed over whether Samuel Langley's Aerodrome or the Wrights' airplane was the first capable of flight.

1926

November 20

Katharine Wright marries Henry J. Haskell

1928

January 31

In response to the Smithsonian controversy, Orville shipped the 1903 Wright airplane to the Science Museum in London, England, as a five-year loan.

1929

February 27

Distinguished Flying Crosses awarded to Orville and Wilbur presented to Orville by Secretary of War Dwight F. Davis.

March 3

Katharine Wright Haskell died of pneumonia in Kansas City.

1932

November 19

Orville attended the dedication ceremony to Kill Devil Hill National Memorial in honor of the Wright brothers of Kitty Hawk.

1938

April 16

Wright brothers their bicycle shop and their home relocated away from Dayton and moved to Greenfield Village in Dearborn, Michigan is dedicated in honor of the Wright brothers.

December 17

Henry Ford, founder of The Ford Motor Company and promoter of the assembly line method for mass-production, was a guest at Orville on Hawthorn Hill in Dayton on the 35th anniversary of the initial flight.

1939

1st December

Wilbur as well as Orville's older brother Lorin Wright passed away in Dayton.

1940

July

1903 Wright airplane displayed on display at the Science Museum in London dismantled and put away in secure storage throughout World War II.

August 19

Wilbur along with Orville Wright Memorial located in Dayton and close to Huffman Prairie was dedicated.

1942

October 24,

Smithsonian Institution published The 1914 Tests of the Langley "Aerodrome," a brochure that apologized for and changed its previous statements regarding the priority to Langley. Langley machine, thereby marking the end of the Smithsonian-Wright dispute.

1944

Orville constructed a cypher machine to allow automated coded messages.

1948

January 30

Orville died from a cardiac arrest in Dayton. NCR bought Hawthorn Hill after Orville's death. The company renovated all the rooms, with the exception of the library, and left the building as a memorial to Orville.

November 22

1903 plane arrived from The 1903 airplane arrived at the Science Museum in London and was transported to the Smithsonian Institution in Washington, D.C. Plane was officially presented to Smithsonian at a celebration on the 17th of December.

2006

NCR donated Hawthorn Hill to The Wright Family Foundation.

2009

March

The U.S. Congress made Hawthorn Hill an integral part of the Dayton Aviation National Historical Park.

2013

June

The Wright Family Foundation transferred ownership of Hawthorn Hill to Dayton History to preserve, maintain and conserve it as a valuable community asset for the future.

I am an enthusiastic however I am not a skeptic because I have personal theories about the right construction method for the flying machine. I want to take advantage of the knowledge that is in the public domain and as much as I can, do my contribution to the next person who will be successful in the end. -- Wilbur Wright

Chapter 5: "A Dream To Fly"

Milton Wright being a bishop was a bishop, he had to move around frequently and the Wright brothers, along with their other five siblings, followed his father everywhere he went, eventually arriving in Dayton the year 1884. Will and Orv are the way they were referred to among their acquaintances and Dayton neighbors adored them and gave them the nickname "the bishops". They didn't realize they live with twin brothers that would become later world-renowned because of their creations. What was the story behind how this idea of flying machines get its start? What was their motivation? It is believed that their father traveled a lot and returned home with something to give the boyslike toys or a chocolate box. While on a trip, their father discovered the toy that would become the spark that ignited in the minds of the young boys. At the time, the nations of the world were holding the notion of flying, and were closely observing and playing with flying machines, but no one was able to taste victory until. However, we are aware that the ultimate triumph was reserved only for two young boys who were only five and nine years old at the time. France was most likely to be the most enthusiastic about flying. A aeronautical engineer by the name of Alphonse Pénaud had made an enormous breakthrough in the field and had developed a model for a helicopter but he wasn't able to establish its feasibility. Milton Wright couldn't be more than a bit more accurate in his selection of an item for his young boys. He bought a helicopter toy which was a miniature replica of Penaud's flying machine made of cork, paper and bamboo and was adorned with the help of a rubber band which twirls its rotating rotor. The boys were thrilled, and had fun playing with it all day until it fell apart. However, the boys didn't scream about the incident They did what was the easiest option to make one of their own, resembling the toy. Milton Wright had sown a seed of a dream, a wish to fly.

Wilbur Wright (1867-1912)

Orville Wright (1871-1948)

The childhood home of the Wright Brothers in Dayton

Redirection is Rejection!

Each of Wilbur Orville and Wilbur Orville were smart, but neither was interested in their studies. And neither of them graduated. While Wilbur was able to complete the four years of high school due to the family's sudden move to Dayton the school was unable to award him his diploma. As a consolation on April 16, 1994 Wilbur received a posthumous the diploma even after he had been deceased and was burial. If he had been alive the date would be his birthday celebration of 127 years.

Both boys participated in sports , and Wilbur was involved they were involved in a tragic accident when playing ice hockey with buddies in 1885. Wilbur is struck hard on his face that causes him to lose the front of his teeth. The traumatic incident, although innocent, affects an 18-year old Wilbur dramatically. He becomes more withdrawn and cautious and confined to his home for the majority of the subsequent years. He even appeared having abandoned his plans to continue his studies at Yale. In the meantime, his mother was suffering from tuberculosis that was tenacious and she died in 1889 .For an untried Wilbur it was a tough time as he fought the loss of his mother and his lack of ambition and motivation. The flames of a long-held dream remained.

In the time, Orville had no plans to pursue his studies further and decides to establish an printing company in this same year. It's not clear what caused the death of his mother that inspired this decision or if it was just a sense of ennui toward books. With the help of his older brother, Orville creates his own company, but fate was not in his favor. The business allows them to continue to grow but it did not take off enough to become an overnight success. There was a brief period when Wilbur was editor, and Orville as publisher of the weekly newspapercalled West Side News. The following year , it was to become an everyday newspaper called The Evening Item. Then, in 1892 the two brothers branched out into the business of bicycles. This was at the time when penny-farthing was replaced with bicycles in the way we use them in the present. In search of a way to profit from this potential the brothers venture into cycling as a business. They founded the Wright Cycle Exchange shop was an repair shop and a sales and service. Soon after, they began producing their own line of bikes. It was an effort to survive, to help fund an ardent dream. The brothers shared a house as well as a bank account, and the desire to fly. They pounded the penny to contribute every penny for dreams they were convinced of. Their business ventures never reached the heights of the newspaper printing business, or their bicycle shop. It was true that it helped them to remain on the right track. However, as the most stale phrase goes, failure is the reverse of a course. If they had found awe-inspiring success and lavishness in their businesses Who can say how long that the entire world might have to wait until it was finally flying? Every time they fell short, they were undoubtedly re-directed to their goal- the invention that would lead man to unimaginable levels.

Predecessors who walked the path

Sky and it's endless expanse has always fascinated the interest of a. The birds that fly effortlessly across the sky are always an inspiring source of inspiration to his imagination, yet he was unable to fly. God's creation had a flaw that required correction and his desire to fly wasn't satisfied-he did not have wings. Man was able to conquer both water and land. He had invented methods to cross the endless oceans , and intercontinental commutes too were made possible. However, the sky presented impossible obstacles. It shouldn't seem like a shock that it wasn't only those Wright brothers who wanted to conquer the skies. There were many nations with their own pioneers, who were continuously seeking to develop a device that could give wings and lift man to incredible highs, just like the birds that fly flying in the air. They were also studying stories of near-perfect but failed attempts to glide and flying. In a newspaper they read and looked at pictures of German engineer Otto Lilienthal who was deeply intrigued by the concept of flying with a pilot and he was a fervent practitioner and tests in this direction but was never able to achieve success. He died in an accident in a glider in 1896, at the age of age 48 and 36 hours following the accident. His final words were incredibly inspirational- sacrifices have to be made and could have given brothers the motivation to fight for their cause, no matter the circumstances.

Karl Wilhelm Otto Lilienthal in 1896

They initially relied on Lilienthal's aeronautical research, but then had to abandon it in favor of the data their own. But Lilienthal's research and data significantly helped in the brother's initial experiments with gliding. Wilbur was believed to have stated that Lilienthal's work was most significant of those scientists who daringly tried to resolve the problem of flying that had kept numerous aspirants slumbering for long. While George Cayley, Percival Spencer, Francis Herbert Wenham, Louis Pierre Mouillard attempted to fly many years before Lilienthal and are to be the greatest aviation pioneers It was Lilienthal that was the very first engineering genius to have his brief flying career to establish and convinced that flying wasn't just an idea, but was a possibility within the next few years. In 1909, just six years after the first flight to fly at Kitty Hawk, Orville while in the course of demonstrating flight in Germany was believed to have visited Lilienthal's widow in a way to honor his unimaginable contributions to aviation industry and the way it helped the pilots in their initial explorations at the beginning.

The year 1896 marked the crucial year in the development of aviation. In 1896, that an American pilot named Samuel Langley was successful in flying a steam-powered airplane, but it was not manned and did not have fixed wings.

Samuel Langley (1834-1906)

Octave Chanute (1832-1010)

Octave Chanute, an American engineer in aviation who was a passionate enthusiast, invited all those who were involved in the construction gliders to visit the beaches in Lake Michigan. In this event, the Wright brothers lost their daughter Lilienthal. Lilienthal passed away. The Lilienthal's death has left a lasting impression on those who knew those who were the Wright brothers. They decided to dedicate their entire lives to the study of aviation research and flight. This is the beginning of their relentless work that leads Wright brothers through Wright brothers through numerous failed attempts and failures before they finally experience victory.

The Chanute-Herring biplane hang glider during an 1896's demonstration at Lake Michigan

While both brothers have presented a single picture and are equally acknowledged for their historical achievements, biographers suggest the fact that it was Wilbur the elder who was the first to initiate the project in the beginning of 1899. The logbooks of Wilbur show that it was usually recorded in the form of a first-person like my plan and machine. Later, it was documented as our. While there is no doubt that the contributions of Orville are no lesser and it would be terribly unfair to compare the brothers' contribution or assess their talents individually It is agreed the fact that Wilbur who was the one who set the pace and, in all likelihood, Orville did not hesitate to be content to follow his lead. Many also suggest that, up until the year 1902 it was Wilbur who was the pilot of the gliders. This could be because he was the older one, and it was largely his idea , or it could be that he was acting as the elder brother who was protecting Orville out of danger since accidents in test flights were commonplace in the era of test-flight. It's a shame that both brothers did not get got married, and neither had any children to carry on their tradition.

The Breakthrough Invention

The early pioneers of the aviation industry like Lilienthal could lift a vehicle using the air-lift concept, however managing the equipment and directing it was an ongoing challenge, even for the brothers. A major leap was made when the brothers developed an apparatus that was three-axis controlled that had its center of gravity as the center in the 3-axis. This allowed pilots to steer the plane, but also keep balance while flying. This was such a brilliant design that it remains the same in all modern fixed-wing aircrafts.

Lilienthal's passing didn't stop the brothers from following his plan. They were aware that if they wanted to resolve this issue completely, they needed to master the technique of controlling the machine before they could even attempt to pilot a motor-driven vehicle they needed to overcome the challenges of maneuvering and steering. The tragic death of Percy Pilcher, a British pilot who died in a crash with a glider, only confirmed the concept. The brothers were aware that the machine flying the three-prong obstacle that included wings, engines, and the controls.

To unravel the mystery of wings they sought help from the masters of the sky, who were adept in flying, as well as in-flight bending and leaning birds. Wilbur was acute in observing how the birds lean their wings to alter direction. It was apparent that the angle near the top of the wings could be altered to move to the left or left. However, how can a human-made wings made of aluminum or steel be bent according to the person's taste and yet maintain an equilibrium in the lateral direction?

The brothers thought about this issue for a while but with no success. One day in his bicycle shop , while bent an extended tubing, Wilbur was convinced that he'd come across a solution to the wing. Thus, the concept of wing-warping was created.

The pioneers of aviation saw flying in the context of other modes of transportation, the only difference being that flying was setting oneself up at an elevated elevation, whereas ships and rails were leveled by water and surface respectively. Additionally, controlling the movements of a pilot seemed alien to them, as they believed it was not possible for a human to control the machine as the wind fluctuated in its direction or force. However, the Wright brothers insisted on having an individual pilot to be able to steer the machine at his own discretion and were aware that the answer to this puzzle would be the first step in the development of flying. This is precisely why the gliders that they created were equipped with drooping wings, known as anhedral wings, which were unstable, unlike the gliders that their predecessors, such as Langley and Chanute created with dihedral wings (slightly downward bent). In 1903, the Wright Flyer was designed with anhedral wings as despite being unstable, the wings were less prone to the gusts which blew through the cross winds.

In 1899, Wilbur was working on putting his theories to the test. He created a model one of a box kite that put wingswarping theories that were put into practice. Biplane kits were created with a wingspan of 1 1/2 meters. The theory appeared to be working because the twisted wing was able to be leaning or slanting to one side , making the possibility of manual control. Inspired by the success of the box kite, the brothers set off for Kitty Hawk in North Carolina to begin their studies with gliders. The location Chanute and Chanute believed to be the most conducive was the mid-Atlantic coastal region since it has sandy beaches that could provide a an easy landing area and less wind gusts that could hinder the taking off. Wilbur has gone to the extreme and had also gathered Weather Bureau data as well. They finally settled on Kitty Hawk that was the closest to the locations that Chanute has suggested. Also, it was away, giving the brothers a space away from the glare of paparazzi and media. It was also far from the attention of paparazzi and media. Lake Michigan demonstration in 1896 was a disaster with media furious and crowding the camp.

Trial and Error

Put Chanute's double-decker hang-glider design, and Sir George Cayley's research that a cambered surface gives greater lift into usage and the Wright brothers developed the first machine. Lilienthal's gliders, which the brothers kept their eyes to had cambered wings, and proved more effective when flying than flat surfaces. However, they made some changes. The earlier designs had a horizontal elevator located at the rear of the vehicle, however the brothers placed it at the front. They believed that this would make their passengers safe from a crash should things go wrong. It was the same nosedive which caused the death of Lilienthal. Another significant difference that gliders had was the tail, or absence of it. Wilbur thought the tail was not needed. However, the first designs didn't have one. Kitty Hawk glides had been narrow and lacked low-quality performance. The glider was partially unpiloted and an unballast was tied to the ground. It flew like the kite. Wilbur was so enthralled by the gliders produced as many as 20 in a single day. The tests were repeated only a few miles away from Kill Devil Hills and their experiments went on unabated for the following three years. The elevator didn't go as smoothly as they expected, however it was successful because there weren't any accidents, or injuries. The decision made to place the elevator at the front was correct.

The glider of the 1900s experimentation

The original designs didn't provide the desired lift, consequently Wilbur created a glider with bigger wingspan. This model was constructed in 1901 was also not without problems, it stopped at least a couple of times and produced just one third of the lift it was designed to in addition to the difficulties in turning and maneuvering. It was always in the opposite direction of a turning. However, the benefit is that forward-facing elevators facilitated a the safe landing of the aircraft, meaning that the tendency to nosedive was stopped. Therefore, all accidents were thwarted. It also strengthened the design of the canard by twisting wings.

The tail-less glider that has its nose pointing skywards, and a wider wingspan

Although there were no accidents or nosedives, Wilbur was crestfallen as the number of calculations or calculations seemed to be adequate enough. The designs they'd been working on for years revealed new weaknesses every time. When they returned home from 1901's experiments Wilbur is believed to have sighed in denial that man would never fly for a thousand years.

A Minor Anomaly

1902 was the year that brought a important leaps in their research. There was a small error within the equation of lift which was a barrier between man and his ambitions to reach the sky.

Despite all the information available, the lift was not enhanced, and they couldn't pinpoint the source of the problem. The brothers were compelled to question Lilienthal's data. The equations the pioneers used before the Wright brothers included an air pressure value of 0.0054 that was known as"the Smeaton coefficient. Lilienthal's equations also utilized the same figure. In actual fact, for hundreds of years, this was the basis of the equation to calculate airlift.

But, Samuel Langley had used an alternative value that was slightly lower. Wilbur wanted to know the exact value and decided to conduct experiments by himself to find the optimal value. Based on the data gathered in his 1901 studies in Kill Devil Hills he arrived with a number near 0.0033 which was more than Langley's. The idea was to increase the lift considerably. Through the wind tunnel they designed, it was possible to calculate the lifting created by the generated air. This study revealed that the data contained mistakes that they needed to address. The numerous tests they ran during the time confirmed that the lack of lift of the gliders of 1900 and 1901 was largely because of an incorrect Smeaton value. This could be the reason why Lilienthal also crashed.

In 1901, Chanute was an aviation enthusiast who was known for his encouragement to the brothers, invited Wilbur to speak at the Western Society of Engineers. It was their first public appearance. they spoke to the public with their ideas. In the conference, a comprehensive report was given on the experiments they conducted and their test flights.

One Hurdle After the Other

The wind tunnel research was an enormous leap ahead. They had made significant advances in their work. It was test-driven on around 200 wings of various designs and shapes to find what would result in the highest lift-to-drag ratio . A detailed test was carried out for 36. The test was one of the most significant and rewarding moment for brothers said a biographer, in particular considering the limited time. Wilbur recognized the significance of wider and longer wings than the larger wings they had relied on until now.

Comparing 1901(left) with 1902(right) gliders

In light of new knowledge gained, in 1902, they developed and built a glider that took into consideration its Smeaton number, the flatter airfoil, and longer narrower wings. A major addition was a vertically steerable fixed rudder which was situated at the rear of the aircraft. The pilotless flight was the first as they had always been. A fascinating aspect to take note of in this regard is that they'd acquired more experience flying and piloting gliders than other pioneers of the present. With bated breath, and a sense of trepidation that was inexplicably high, they took their first flight without a pilot in those same dunes where they were in last year and they flew over the Kill Devil Hills. They were relieved to discover that the air was much more affluent. Finally, they can taste a little of satisfaction.

The rear rudder that was steerable eliminated the yaw-contradiction (the first design in which the it turned to the opposite of the direction it was intended to go) that hampered their prior design. However, it was not yet the enough to declare that the brothers had made the first in history. By removing the adverse yaw with such efficiency that it created a new challenge. The obstacle now had an entirely new name: well-digging.

The brothers finally get the fruits of their labor in 1902 when they introduced a the rear rudder that could be steerable.

Orville believed that the root of the issue could be an rudder that was fixed. A moving rudder was an likely solution and this is precisely what the brothers turned to. The brothers were referred to the incredible idea of moving rudders. Its primary function was not to turn, but to help align the aircraft when turning.

Wilbur making a turn using an movable rudder, and the design of wing warping in 1902

From 19 September until 24 October, the brothers performed up to 700 to 1000 glides. The longest flight ran for 26 seconds before gaining 622.5 feet, or 189.7 meters. It was so exhilarating and satisfying that they immediately realized they were ready to take on an electric glider.

The control system with three axes was created to address the problems of control and wings. The third component was the engine. It was impossible for any company to manufacture an engine that was lightweight but powerful enough to give lift. The brothers realized they had to build one on their own.

On the day that marked the beginning of history, 23 March 1903in which the twins applied for a patent on their aircraft. It was only in 1903 that they develop the powered flyer and the glider of 1902 is widely regarded as the defining moment in the invention of the airplane.

Chapter 6: A History Of The First Powered

Airplane

Wright Flyer from 1903 Orville Flying, Wilbur at the wingtip

The brothers were ready to construct their own powered glider. They used the wood Spruance which was sturdy and durable, yet light an ideal choice for the glider. The coverings on the surface were made using mullin. The propellers were made of wood and made by hand. But, the efficiency of it wasn't up to the level. Wilbur wrote in his logbook from March 1903 that the engine was 66 percent efficient. The requirements for engines also were to be satisfied. After trying to contact a myriad of engine makers, they could not meet their requirements and they went to their mechanic. Within less than two years Charlie Taylor built them an engine. To keep the weight of the engine down, they utilized aluminum, which was a rare commodity in that time. The propeller was powered with a chain that resembled an old bicycle. The Flyer was designed to be cost-effective and cost less than a thousand dollars. It was a tiny drop in comparison to over fifty thousand dollars used by the government to finance Langley Aerodrome. In just a few weeks of the initial powered flight, the Ohio newspaper published stories that sang songs of praise for the Wright brothers' achievements. The newspaper proclaimed the dawn of a new era in which flying was a fact and not be a mere an esoteric goal. The newspaper emphasized how the flyer of the brother eclipsed the dirigible balloons created from French engineers.

14 December 1903 First Powered Flight

The date chosen to fly the pilot's first flight was delayed because of the weather's turbulent conditions. The 13th day of December in 1903, was Saturday and that's why the following day Wilbur chose to go for the flight. It was the anniversary of the first hot balloon test flight, which was carried out by Montgolfier Brothers. In 121 years ago, on the 13th of December in 1782, the Montgolfier brothers tested their hot balloon for the first time. There were other hiccups in the road to success. A minor accident caused damage to the shaft, and needed to be fixed. After the necessary repairs, on December 17, it is again being tested.

An unforgiving end to an initial Flyer with ambition

On 17 December 1903 at 10:35 a.m., Orville made his first flight test, flying to 120 feet in only 12 minutes, with the velocity of 10.9km/hr. Wilbur was next to test it, and then Orville once more, flying to heights of 175 feet and 200 feet, respectively. About midday, Wilbur was the one to take last ride. The last ride was turbulent and almost fell into the ground. The frame was cracked and needed to be repaired. When the brothers tried powered flights, there were five people who gathered to watch the amazing feat. The men who were there assisted Wilbur to pull the Flyer to the ground following his final flight. However, a fierce breeze swept in on Wilbur's Wright Flyer was carried away by the wind, which was too strong and it was flipped around many times before the pilots could get control. The flying machine was badly damaged and it was not able to fly again.

The plane was then shipped into their house in Dayton. A few years later Orville would fix the wings that were damaged and then it was donated to the British Museum. It is now in the Smithsonian Institution in Washington D.C.

Inadequate acknowledgement

After many hours of endless trials and tests, the brothers finally found their first victory. The road bumps did not deter their determination or affected their determination. However, even after achieving the feat of flying an electric flying machine, they weren't recognized or appreciated initially. Brothers had sent message to their father announcing the flight's success in order to tell the media. However, Dayton Journal Dayton Journal declined their news and stated that the flight was not long enough to be significant. The success story, however, was leaked, and the next day an untrue story was created around their triumph and printed in newspapers across the nation including the hometown of Dayton.

In the month of January, the brothers made the decision to go publicly. However, to their dismay, the announcement didn't generate an enthusiastic reaction. The tale of this amazing victory was not well-known and the story soon died as a depressing end.

France was always interested in aviation, nonetheless noticed the Wright brothers' flying successes and was thrilled to see the Wright brothers.

Going Public

Despite all the accomplishments in 1903 the federal government appeared to be that it was not interested in the brothers' accomplishments since they didn't have many mouths. They wanted to dedicate their lives to the cause they were made to fulfill. Even though Langley and Maxim the aviation pioneers received huge financial support but the Wright brothers were not wealthy and did not have any financial aid to pay their bills. If they had to shut the bicycle shop, they knew that the aircraft will provide them with butter and bread. They were unable to afford to sell their invention, which could have led the brothers to a state of poverty. So they needed to keep their inventions confidential and secure.

To cut costs, they decided to try the Wright Flyer II at a cattle ranch within Huffman Prairie. They were delighted to learn that the ranch owner didn't have to pay any money. On May 23, they let down their guards and allowed the media to observe their experiments, but made a deal that no pictures were to be taken. The attempt ended up being a mediocre occasion as the lack of wind and engine problems did not allow for a solid launch. The media were shocked and for the following year and half , they put their heads down at the two brothers as they sat in their aircraft.

It is not clear if the brothers intentionally staged an unprofessional performance in order to get media off their tracks. They continued their experiments, without distractions from the annoyance of journalists. The unpredictability of the terrain and the fluctuations of the wind brought new challenges such as instability and landing issues more in contrast to those of the Kitty Hawk experiments. The landings were messy and chaotic leading to hard landings, and minor injuries. These setbacks didn't stop their flight. One particular day saw Wilbur took off at 400 meters in height far more than their previous best chance in Kitty Hawk. To counteract the low lifting capacity, they chose to make use of a weight which could launch the lift. The 20th September 1904 was the day that Wilbur was able to complete a circle that covered 1244 meters in just a minute and a half. Wilbur therefore made history by claiming as the first person to fly a machine which was more powerful than air. In 1904, Wilbur and his brothers had completed 50 minutes in the air.

In 1905, the brothers developed and built their third flying machine and named it Wright Flyer III. The flight test proved to be disappointing because the third model did not perform like Wilbur was expecting. One of the tests, Orville experienced a serious crash and Wilbur realized that the machine also needed massive adjustments. He constructed a bigger forward elevator, and the rear rudder was increased too. They also ensured that all three axes i.e. the roll, the pitch and yaw could be independently controlled. This greatly improved quality of the control and stability. The brothers were able to fly for more than 30 minutes and travel over 39 kms.

Flyer III Orville Flyer III, flying for an record-setting 33 minutes, 17 second

The flight's final trial was led by Wilbur and their beloved friends and their father Milton was also present. The audience gathered was able to watch the Wright brothers as they took on the skies and flew in circles across the sky of azure in Huffman Prairie. After having a look at the historic flight that were made by the Wright brothers, who had rejected the year before The reporters met one day later, but the Wright brothers weren't ready to fly to the media.

The success of Flyer III proved to them that they had made an aeroplane that was able to be used in a practical way. The majority of the photographs taken during 1904-1905 were destroyed in the Great Dayton Flood of 1914.

The Flyers, or the Liars! !

The sight is the word that will be believed by the majority of us. It was believed that the Huffman Prairie flight was witnessed only by Wright's generation. There was no one outside who seen it, not least of any of the mainstream media. There was plenty of speculation, and doubts were widespread. The magazine even said that the brothers were as liars. The article published in Herald Tribune carried the headlines either flyers or liars. Their flight was not successful enough to take over the headlines of local newspapers. Their success wasn't celebrated nor did it bring their name to the forefront of media. In reality, they were not celebrated. The Dayton Journal had published their triumph in the context of business and agriculture news, without making it appear significant. Nobody believed it and they had dismissed it as a as a hoax.

But the disbelief that was widely spread was not without a basis. There was no evidence to support the event. The brothers also share some of the blame since they kept the incident secret for good reason. The media and reporters were unaware of anything about the tests. Local newspapers were cautious about publishing content that had no reason to believe in its authenticity. In the absence of the controversy caused by media, external media also were hesitant.

When the news came out and the world began to talk about it, the brothers were unable to fly until they signed a agreement in the black and white. They contacted the governments of the US, France, Britain and Germany and offered to sell their invention, but were disappointed when they did not show the invention before signing the agreement. The American government's concerns were understandable given that the nation's most famous scientist Langley's aerodrome was heavily financed ($50,000) but only to see it sink in the Potomac River, twice. As the Langley brothers sat with their famed aeroplane lost to history while other aviation pioneers basked in the spotlight and received praise.

European Media make a mockery of the brothers

The anti-Wright brothers ' plight was embraced by European aviation became more and more popular in the year 1906. France newspapers even slammed the brothers as Bluffers. This was a time of immense pain and suffering for the brothers.

Ernest Archdeacon, a French lawyer who was also an aviation enthusiast, wrote articles about the brothers. They were an enraged attack and completely mocking. He also made bold assertions that the French were the first to create an aircraft that had practical use. The European media had also been averse to the brothers. However, they did not let Ernst be aware that within two years the brothers would receive praise and approval from Ernst.

In 1908, they flew their first flight to France and Ernst was able to see his own reflections that he had committed a serious error to the brothers by smear-bashing them in 1906.

In the following months, brothers continued trying to get an official agreement and contract but with no success. They also continued to test different tests, including one that involved a launch from the ocean, but this did not work neither. American government's indifference to them urged them to attempt the waters in France. They French were always eager to take part to conduct aviation-related tests. In 1908 they signed their first deal with an French company. In the dark until that point the brothers began experiencing an end to the long and dark tunnel. At this point, the aeronautical wing of the US military finally agreed to an individual-to-person demonstration. They were clearly impressed at by this time.

The agreement that they signed with the French company required a person to fly alongside the pilot. The brothers return to Kitty Hawk and do the adjustments required. They also work rigorously, honing their control and piloting skills to ensure that their public demonstration doesn't get out of control. However, both brothers enlisted the helper to test flying with a person since they had promised their father that they wouldn't fly in tandem. One of the flights Wilbur lost control as he was still wet in his ears, and only starting to get comfortable in the controls with a person on. However, the flight tests were stopped because the aircraft was damaged and destroyed.

Flying in Public

They finally got the interest they had always had hoped for. The contracts to be signed only if the type of flying that met certain conditions. They were flooded with test flights and demonstration requests, so they chose to share the task between themselves. While Wilbur was flying from Europe, Orville would demonstrate flight operations in Washington.

After receiving a lot of mockery and ridicule by French journalists, Wilbur knew he had to prove his shrewdness. The aviation community had created an admonitions of their invention and had branded them bluffers. On August 8, 1908, Wilbur flew for a one minute and 45 seconds, but his twists and turns using his machine that it made people watching him with jaws that were clenched from the ground. The crowd almost fell over as he took technically difficult turns like flying in the form of the number 8. This was the most outrageous flying show and the crowd who were there knew Wilbur was writing his history. The reaction of the crowd was thrilling and thousands of people had come to see history being written by a boy who was slammed and labeled an obnoxious bluffer by the media couple of years ago. The brothers made headlines and became world famous. The same newspapers that slammed and published scathing reports about them, sang praise songs about their feats of courage. The Archdeacons were quick fix his past mistakes. He was open to their accomplishment.

On the 7th of October 1908 Wilbur flew along with the spouse of the business representative, they were the very first American woman to fly alongside the brothers. Another person by the name of Leon Bolle, a French auto manufacturer, also flew along with Wilbur for his passengers. It was in Bolle's manufacturing facility that Wilbur assembled his Flyer.

Orville was also reaping the rewards in America. He was showing flight techniques for his fellow soldiers of the United States Army at Fort Myer. On the day that made history, 9 September 1908

Orville flew for record-breaking of 62 minutes and 15 seconds, and left the crowd awestruck by his skills in piloting and the stability that the machine displayed. However, a tragic event was waiting for the brothers, destroying their self-made pride.

Fort Myer Crash, 17 Sep 1908.

On September 17, 1908 Orville embarks on a flight alongside Army Lieutenant Thomas Selfridge. It was a defining day. Selfridge was selected as an official witness. After a successful takeoff just a few minutes into the flight, the propellers exploded and caused the plane to crash down from 100 feet in height. Selfridge suffers a fatal skull fracture and passes away that same evening. The plane crashed was fatal for the very first time in history of man and Selfridge was the first human to die from the result of a plane crash. Orville was also seriously injured with a fractured left leg and a few fractured ribs. A decade after the accident, because of constant discomfort, Orville is hospitalized . X-rays found bone fractures as well as dislocated hip.

In fact, this incident just made their resolve more steadfast. Wilbur broke records in both length and height. Orville was with him the following year after a painful recovery. By 1909 , the brothers were much more or less celebrities, and were adored by both the public and media. The royalty were not any different and their interest was piqued and the queens and kings from Spain, UK and Italy were eager to watch Wilbur take off to new heights. Wilbur's flying success gained him so much attention that he started taking press officers and other officials of the state as passengers.

Wilbur showing his flight skills in 1908

Wilbur was able to train two pilots and also sold his first aircraft to a French company in accordance with the agreement. Then, in April of 1909 Wilbur built another Flyer and began training more pilots as well as giving more demonstrations. Federico Valle, an Italian cameraman was the first one to take off and record a motion image mounted on an aircraft.

Even though it's late, Success finally arrives at the doorstep.

After a stunning show in Europe the brothers make their way back to America. William Howard Taft, the 27th president of America had invited the brothers to the White House. In July, the brothers completed the two-seater airplane which could fly at speeds of 40 miles per hour and landing safely. The plane was purchased by the Aeronautical Division of US Army for 30,000 dollars. The speed requirements were achieved and the brothers finally receiving recognition and rewards for their hard effort. They were also awarded $5000 to surpass the speed limits.

1909 brought about a significant triumph for Wilbur and his brothers. Wilbur performed a staged circle around the Liberty statue and then took off on a spectacular flying flight over the Hudson River that lasted for 33 minutes. It was a goose bump-inducing show that was watched by over a million New York dwellers. If this doesn't strengthen the Wright brothers' fame What do you think it would?

The Third Wright

There was a third Wright brother who rose to fame in the latter the 1900s. The younger sister of the brothers, Katherine Wright, who was the one who worked on their Dayton home as the boys were spending a significant portion of their time in the bicycle shop, polishing their dream they had been working on since their early childhood. Following an accident in 1908, Fort Myer crash in 1908, Orville was nursed back to his health by Katherine. Katherine's sister had been with her to the White House and on February 1909, Wilbur took her to the White House in his flyer.

On May 25, 1910, the brothers flew in tandem for the first time. This was the first moment that the two wrights could sit close to each other as they pursued their mutual desire to fly. After he had a great time taking his older brother Wilbur to a spin it was the 82-year-old of father's time to experience the thrill of flying. Orville was able to take his father on a flight that was 350 feet high. There is a legend that the ecstatic Milton Wright was heard shouting to take off further and farther. The first occasion that Milton Wright flew on a plane.

Legal War and Wilbur's Untimely Death

On the 22nd of May 1906, the brothers were awarded a US patent. The patent was a cover for methods for wing-warping, as well as controlling and piloting a machine flying powered or unpowered and steerable rudders, which prevent an adverse yaw as well as a forward elevator designs. A lawsuit of acrimony begins when Glenn Curtiss utilized ailerons, and imitated Wrights design, without giving them the credit they deserve. On July 4th 1908 Curtiss daringly flies an American aircraft June bug. This was an utter wrong and an infringement of a patent that legally belonged to brothers. In 1909, despite of their disclaimer, Curtiss sold an airplane with Ailerons to the Aeronautic Society in New York. A lawsuit is brought not just on Curtiss but also on a number of other pilots, including the well-known Louis Paulhan. The legal battle lasted until 1917, when the patent was expired in spite of being the French being more accommodating to the Wrights. The Wright brothers eventually won this case over Curtiss in 1913, when the judge had decided that the ailerons fell under the scope of patent.

However, Wilbur did not survive the persistent typhoid infection, but he did witness the legal victory after he succumbed to his disease in 1912. The legal saga began to take its toll on their development of new designs, and, by 1911, the Wright's designs were unable to compete with the Europe manufacturers of designs. In the battle to satisfy attorneys and prevail in the fight against Curtiss, Wilbur had to travel a lot in addition to Katherine and Orville think this was one of the reasons that led to disastrous consequences for his health. This war over patents destroyed Wilbur's airplane mission to the extent that, when America began World War I, they needed to defend themselves against French aircrafts.

Curtiss and Wright Tug of War

Curtiss Company had challenged 1913 verdict and sought an appeal. However, in 1914, the appellate court refused to hear the earlier judgment. Orville who was without the protection from an older brother, was pleased to be able to hear the verdict to his advantage. It was a solitary fight without Wilbur to support him. A sad Orville lost his vigor and passion. He decided to stop fighting against monopolies in manufacturing. In the meantime, US had entered war was under pressure to make more aircrafts and asked the highest levels of the aviation industry to establish an association. It was called the Manufacturers Aircraft Association. The member nations were required to pay an amount to all patent holders which included Wright brothers. Wright brothers (though at present it was only Orville). The Curtiss Company and the Wright brothers company received a whopping $2 million. In a twist that was ironic of the 1920 events both companies who were locked in the legal dispute merged under one single entity Curtiss-Wright Corporation and though it might sound odd, it's been a highly successful merger, because the company is still an elite manufacturing firms that design and manufacture premium products for aviation.

Legal lawsuits have ruined their image. They were criticized as untrustworthy and fraudulent. The first fury of their work led to speculations rampant. However, in their defense, some claim that it was just natural for them to seek an appropriate reward and recognition for the many years of work they had dedicated in this direction. The brothers devoted their youth to the cause and were so committed that neither had children or got married.

The brothers have received such criticism or criticisms that obscures their true efforts to make the dream of flying machines reality. The strong bond and friendship they formed with Octave Chanute started to fade as claims and credit were discussed. Ten years of friendship was ended in the end when Chanute ridiculed the brothers.

Wright Company

Established on 1909, November 22nd. The Wright Company had Wilbur as the president, and Orville as vice-president. The patents were transferred to the company at a price of 100,000 dollars. The brothers earned 10% royalties on each plane sold. A factory was built in their home town of Dayton and a test flight field was constructed at Huffman Prairie.

The design of the flyer was altered, and models A as well as model B designed. There was no time when flying machines enjoyed massive sales, so the brothers were required to teach a group of pilots with a salary who would display their skills and win cash prizes. Wilbur did not like it initially, as he thought that it was a shame. But they had to earn a living to cover their costs. The show lasted for one year, and in November, the brothers broke up the group because two pilots lost their lives in the course of the event.

The first air cargo was delivered on the 7th of November, 1910 when silk was flown over 65 miles, from Dayton in Ohio to Columbus, Ohio. Brothers received 5000 dollars for the cost. This was more of an advert than a shipment. They were able to complete the 105-kilometer stretch in just over an hour and six minutes. The silk was strapped to the passengers' seats. It is fascinating to know that silk straps were torn into pieces and was made into souvenirs as a memento from the an air cargo that was the first in the world.

The year 1912 was not particularly favorable for the brothers. Following the loss of Wilbur in 1912,

Orville was distraught and devastated. In addition it was that the aircrafts the US Army had purchased were crash-bombing one after the next and leading to the deaths of pilots. These frequent accidents started to raise doubts about the security and the safety of Wright aircraft. The number of deaths claimed was steadily increasing to 11 in 1913. The C model aircraft was nosedived to the ground and fell to the ground. Orville was now an unmanned fighter. He argued that crashes weren't due to an ineffective design, but rather because pilots were not sufficiently experienced. He agreed with the army on the installation of an indicator to prevent sharp turns. An investigation by the government was conducted and it was found the Type C was unfit for flying, and the American military decided to stop using. The government believed that the engine located at the rear also meant that a engines that nose-dive could smash the pilot, causing death.

Langley Feud

Samuel Langley was the secretary of the Smithsonian Institution. The man had committed his entire life to aviation and had been testing for a long time. But his machines had proven successfully used in unmanned flights in 1896, and another success in 1903. But his highly-anticipated Langley Aerodrome powered by motors that the federal government invested in, was a total failure. It was marred by a tragic crash. However, the Smithsonian Institution showcased Langley's aircraft as the first heavy than air vehicle that was capable of powered and manned flight. This was a grave violation and unfairly put Wright's Wrights machine to the second place. Orville is furious and that triggers the legal battle between the two pilots. In the end, it appears that the Smithsonian organization was in a pact with Curtiss. Since Curtiss was in a lawsuit with his fellow flight enthusiasts, the Wright Brothers, the latter thought it appropriate to cooperate with Langley's unscrupulous claims to smear Orville and damage the Wright brothers reputation. Orville was accused by Smithsonian of concealing the history of the flying machines. He donated Kitty Hawk's Kitty Hawk flyer to the London Science Museum in 1928 and refuses to loan it out to Smithsonian. A few years later, in

1942 the museum finally resigns and also retracts its false claims it made regarding the 1914 tests. The story goes that Orville asked for the return of the aircraft at the British Museum but he was never to see the machine because it was returned to its home after Orville's death, 35 years later than Wilbur's.

The Beacon that inspires and awes.

Following the death of Wilbur It was a lost reason for Orville. He was lonely and devoid of love. Wilbur took his last flight, in the month of June 1911, in Germany. His final days were always on the move for legal actions and patent war, and business matters with The Wright Company. Wilbur was devastated by the fact that years of hard work were forced to be fought over in such unfair ways. Orville recalls how his brother returned from travels white and pale.

The brothers shared dreams of building a larger, more grand home. They named Hawthorn Hill. Hawthorn Hill. The construction was already underway in Dayton. There were Orville as well as

Katherine who looked after the construction work while Wilbur was traveling in Europe. Wilbur could never witness the home. He had one request that he had an bathroom and a bedroom for himself. The home was built however Wilbur was dead two years earlier.

After the death of Wilbur Orville who, like Wilbur was not a keen interest in business , decides to sell the business. After Wilbur's death, three years later In 1915, Orville sells the Wright Company. It is then renamed the Wright-Martin company in 1916.

www.ingramcontent.com/pod-product-compliance
Lightning Source LLC
Chambersburg PA
CBHW050412120526
44590CB00015B/1929